MW00836809

BREEDING THE VAMPIRE AND OTHER CRABS

(BRACHYURA AND ANOMURA IN CAPTIVITY)

HUSBANDRY, REPRODUCTION, BIOLOGY, AND DIVERSITY

Orin McMonigle

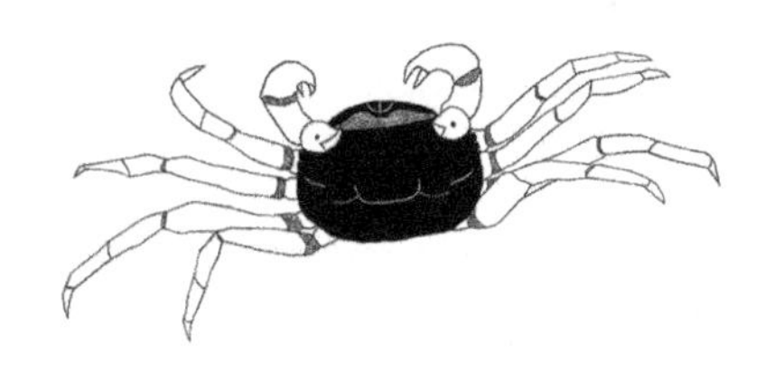
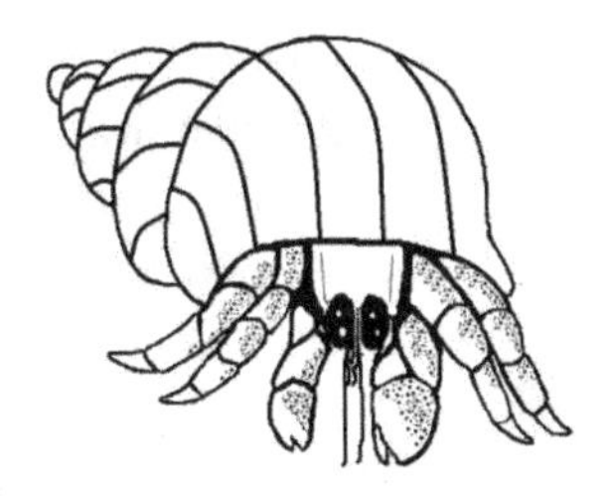
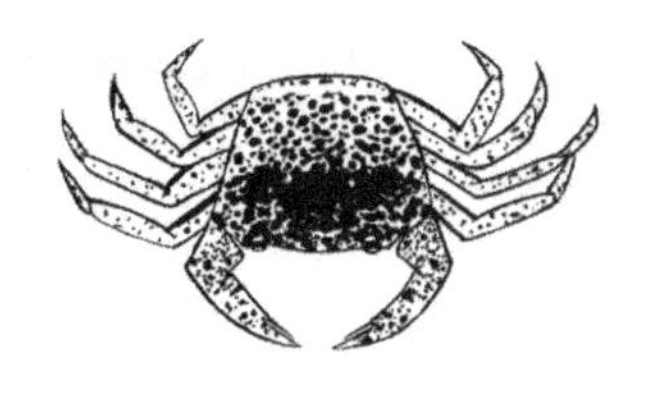

COACHWHIP PUBLICATIONS
GREENVILLE, OHIO

Dedication

I dedicate this book to my father Dan McMonigle. When I was small, he took me to Sea World of Ohio and a dozen public aquariums, including the Cleveland Aquarium, Shedd, and Baltimore. There were fantastic long-armed crabs in the touch pools and marine crabs of various shapes and sizes on display. My father drove me to pet stores where he bought my first pet fiddler crabs, terrestrial hermits, decorator crabs, and sponge crabs. To this day he supports my fascination with crabs. He has waited at my house for boxes of crabs when I was at work, and lets me grab snails and daphnia from his rearing vats to feed my crabs.

Acknowledgements

I would like to thank Christian Schwarz and Annie Lastar for editorial comments and suggestions. Thanks also to Joseph Reich for suggestions and for sharing his experiences rearing and breeding *Geosesarma* species. Thanks to Amber Craigie for comments on the anomuran list. I appreciate my brother Ryan who located littoral crabs for my studies. I thank my wife Sylvia for her endless support. Thanks to my daughters Kree for a great crab drawing and Gwynevere for her temporary interest in various fantastic creatures. Thanks, of course, to God for creating a spectacular variety of crustaceans, specifically the marvelous crabs.

Breeding the Vampire and Other Crabs

Coachwhip Publications // CoachwhipBooks.com

ISBN 1-61646-429-1
ISBN-13 978-1-61646-429-5

CONTENTS

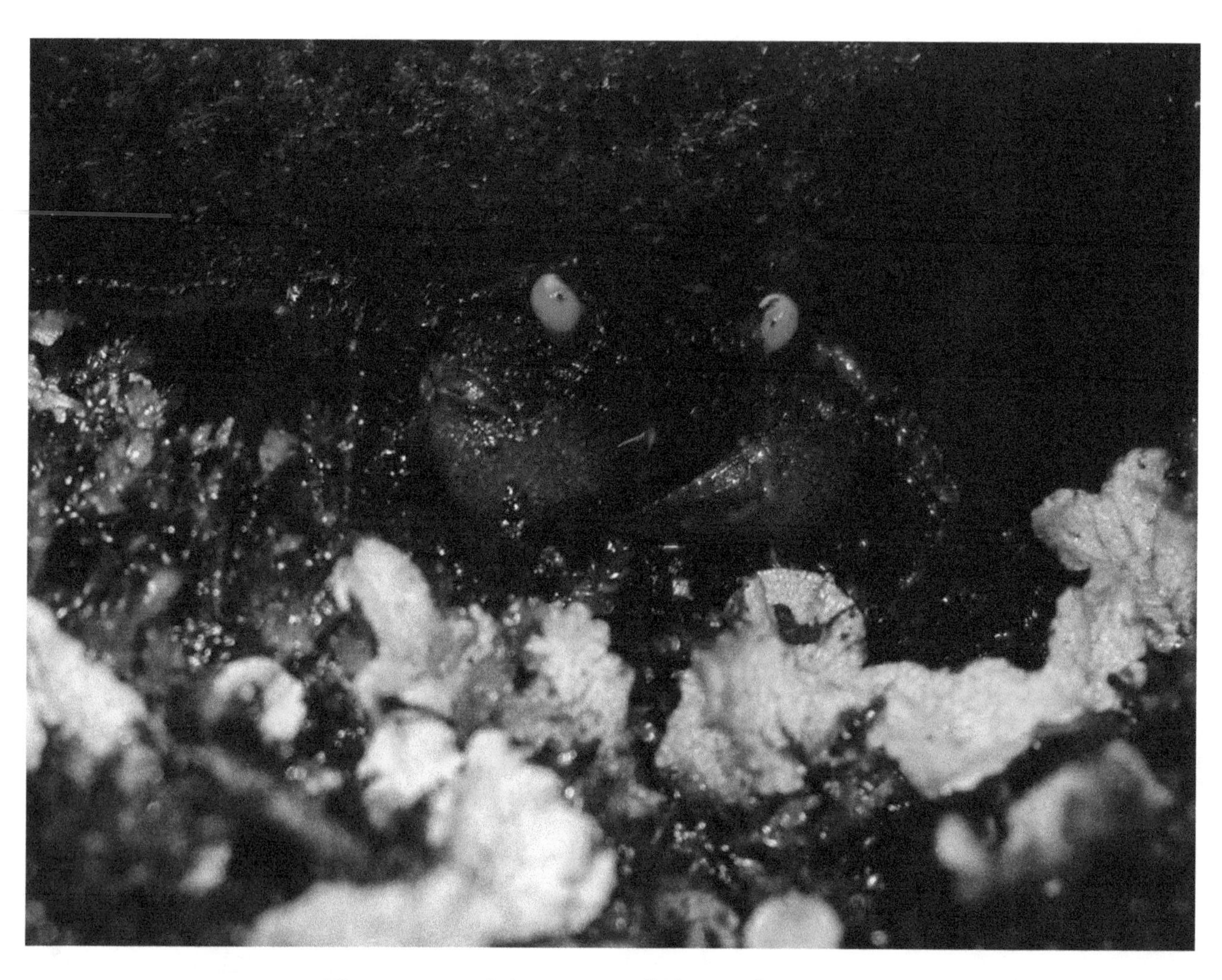

The mesmerizing eyes of the male vampire

INTRODUCTION

Crabs are arguably the most widely kept group of pet invertebrates. Crab enthusiasts are found among various factions of otherwise isolated pet keepers. True crabs (Infraorder Brachyura) and their relatives are integral to the marine aquarium hobby as both clean-up crews and display animals in their own right. Common fiddler crabs are regularly seen in freshwater-only pet shops. Mangrove, vampire and panther crabs make it to freshwater shops from time to time. Larger terrestrial species, specifically moon crabs, are found at reptile shows as small terrarium animals. When anomuran crabs (Infraorder Anomura) are included, crabs are sold in more shops than any other single creature. Terrestrial hermits are available from thousands of pet shops as well as countless curio and tourist shops in coastal areas. Other anomurans, mostly marine hermit crabs and porcelain crabs, are normal fare at saltwater pet shops. (Squat lobsters are very rarely available.) Although crabs are widely kept, literature on captive husbandry and reproduction is scarce, and widely scattered. No other group of creatures is broken up into separate avocations (and separate literature) to such an extreme. *Breeding the Vampire and Other Crabs* is a compendium designed to look at these widely loved creatures as a whole. This text is centered around reproductive biology of the fantastic vampire crabs and other true crabs. Requirements for keeping hermit crabs and other commonly maintained anomurans are discussed.

Available literature on pet crabs is limited to the rare article or short manual, a few pages in marine biology texts and general invertebrate

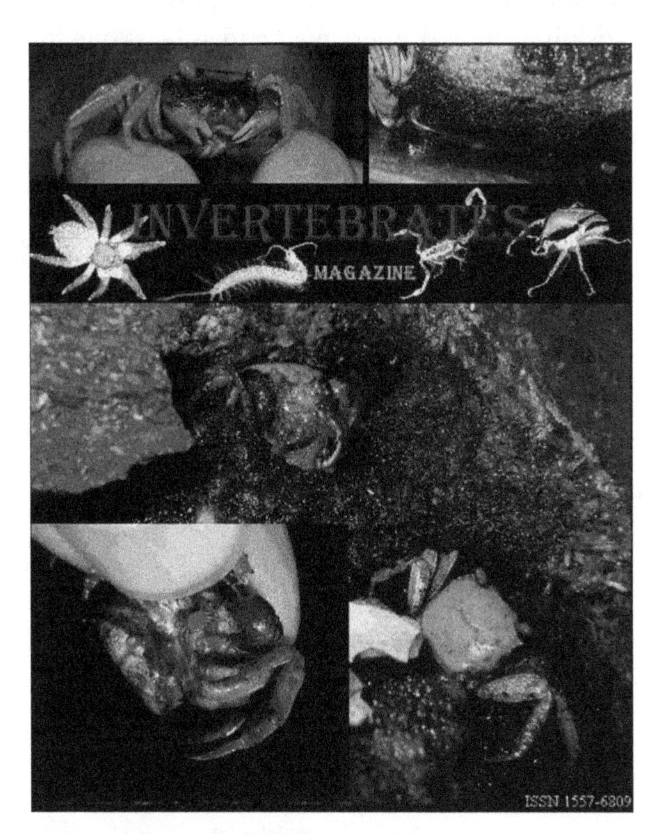

Crab articles have been rare in invertebrate and aquarium magazines, while covers are practically unheard of.

guides, or to children's books—little of which caters to the serious crab enthusiast. A handful of husbandry articles have been published in the German *Arthropoda, Reptilia, and DATZ*, the Danish *Exotiske Insekter*, and the English *Invertebrates-Magazine*. There are also the rare articles in aquarium magazines such as *Tropical Fish Hobbyist*. A small handbook devoted to terrarium crabs (Rademacher and Mengedoht 2011) was published a few years back, but it is unfortunately only available in German. More recently, a longer, small-format paperback specific to the genus *Geosesarma* (Höhle and Singheiser 2016) was published in German and translated to English. There are many dozens of small books on keeping hermit crabs (Nash 1976, Pronek, 1982, Vosjoli 1999, Fox 2000, etc.) and at least one for fiddler crabs (Halton 2013). Brief inclusions in general marine texts like *Marine Atlas 1* (Debelius & Baensch 1994) and *Encyclopedia of Marine Invertebrates* (Walls 1982), or in care guides like *Invertebrates for Exhibition* (McMonigle 2011) include sometimes interesting, but quite limited information. The most numerous literature is probably the many dozens of children's books on hermits, including a few for true crabs, that relay short life lessons or offer some very basic biological details for readers ages 5-9. *Breeding the Vampire and Other Crabs* brings details together from scarce resources and includes many first-hand experiences from the author. There has never been anything like this text, which provides a more comprehensive overview of crabs kept in terraria and aquaria for the hobbyist.

When it comes to husbandry suggestions, an important question is whether or not the author has an extensive background in keeping the animals discussed. Has the author kept species over extended periods or through multiple generations? My first paper (McMonigle 1989) entailed keeping marine crabs, showing my lengthy commitment to this particular passion. I have reared *Geosesarma* through multiple, consecutive generations and have spawned dozens of marine and brackish water crabs. I have kept many species of Anomura and Brachyura throughout what appears to have been their full natural life cycles.

When writing about a group of invertebrates, I do not even try without a flagship species or genus that is available, beautiful, and readily bred in captivity. This book is meant to document and promote captive husbandry of *Geosesarma,* but it includes my experiences keeping different pet crabs over the decades. The idea for a book began in 2010 with *Geosesarma* rearing and breeding successes, but I did not begin to work on the text in earnest until 2013. The plan was to have it published by winter of 2015, but I kept encountering species to which I had never had access before. I also discovered I did not know as much as I thought I knew—despite decades of maintaining crabs in artificial habitats. I hope to detail what I have learned through both success and failure, to help other keepers better maintain their crabs. I am amazed that keeping a species for six months, a few years, or even decades can distill down to a few sentences. I offer my experiences of failure and success as a guide, but there are many more ways to better care for or harm them than I will ever know. Hopefully this text will inspire others to discover and record husbandry data about the species discussed within, and the new species sure to come.

Geosesarma (vampire) crabs are the first widely available crab to give the average enthusiast the ability to breed and care for successive generations in captivity with relative ease. Almost any crab species can be an enjoyable pet, but planktonic larvae restricts availability of most species to wild-caught animals. This prevents the enjoyment of observing life cycles and offspring in captivity. Other terrestrial and freshwater crab genera offer similar opportunity for captive breeding to the general enthusiast,

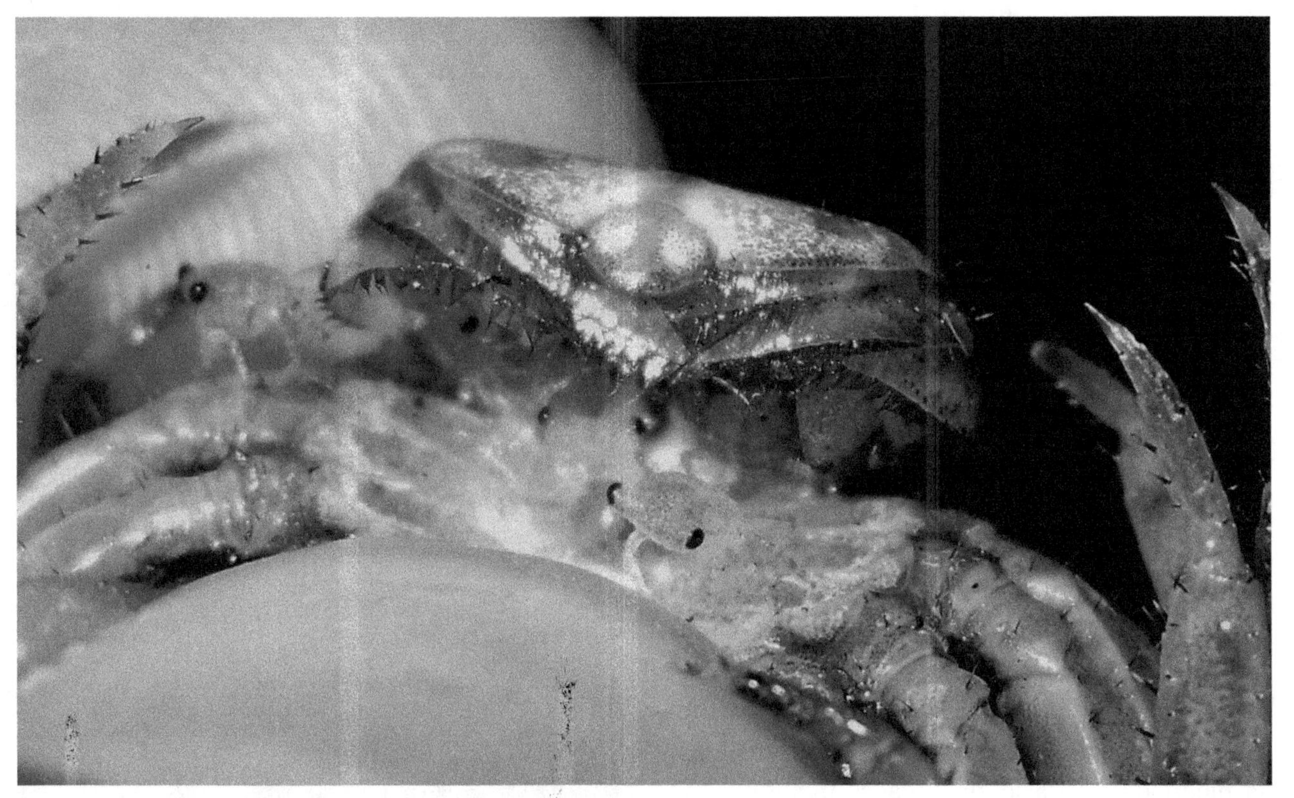

Hatchling vampire crabs, the beginning of captive breeding

but they have been rarely available in the North American hobby. *Geosesarma* spp. are also featured because they display fantastic colors, are energetic, and are fun to watch. They have proven to be spectacular terrarium pets. They are not nearly as destructive to their artificial enclosures as most other crabs. They do not require large spaces or special filtration. Lastly, I have more details for *Geosesarma* than other genera because I have successfully bred successive generations and multiple species.

Common names for crab species seem to be less consistent than for almost anything else. This is partly because many distantly related crabs look like each other and partly because a single species can vary widely in color, size, and shape. Sexually mature crabs can differ greatly in size according to age. Color and shape can also be unique at various stages of an animal's life. It is often difficult to know what part of the world pet crabs came from, let alone determine a specific country, city, or locality. Something sold as a "Peruvian" crab might well have been collected in Africa. One of the longer trade names is the "electric blue Halloween hermit crab." Long names are not the problem. The greatest difficulty is consistency. Names like "red crab" are used for dozens to hundreds of species from different families or even crustacean orders. Scientific names are certainly more useful and readily assigned than common names, however many identifications in the hobbyist trade are just guesses that may be entirely wrong. This text will follow the most prevalent common name in current usage. There is no standardized name list to draw from (unlike standard name lists for tarantulas, cockroaches, and a few other invertebrate groups).

Terminology for crustacean biology can be difficult to follow, especially for the new enthusiast. Even advanced keepers may have a difficult time since different types of crustaceans can follow different terminology. Sometimes this is because the initial hatchling is different in structure: brine shrimp, triops, and most other crustaceans hatch out as nauplii, but malacostracans hatch out with well-defined body segments (as zoea). In other cases, the difference in terminology reflects only specified jargon. True crabs transform from zoea into megalopa while hermit crabs also start as zoea, but become glaucothoe. The same stage (after zoea) is parva in shrimp, while isopods hatch as mancae (post-larval juveniles). Names used for early development are further complicated because the same stages vary in the number of instars and specifics of the body form. The body parts of crustaceans often follow common names for analogous body parts of arachnids and insects, but some crustacean terminology and body structure differ greatly. There are unique terms such as pleopods (thoracic appendages including the walking legs), pleuropods (abdominal appendages or swimmerets), chelipeds (front legs usually ending in a claw), and maxillipeds (leg pairs adapted into mouthparts). Specific crustacean terminology is necessary for some explanations in this book, but nontechnical words will be used as much as possible.

True crabs have been increasing in popularity as pets. This is *Gecarcinus quadratus* at a chain pet store in 2017.

This text caters to the general crab enthusiast. Hopefully the information provided will allow new and seasoned keepers to better maintain their animals. It is hoped the pictures and species accounts will entice readers to acquire and keep a wider cross-section of these beautiful and personable invertebrates.

MAN AND CRAB

A person who studies crabs is a carcinologist. This comes from the Greek word for crab (*karkinos*). The Latin word *cancer* may be the scariest word in the English language, but it originally was the name of the creature we call a crab. The unchecked growth of malformed cells with its large veins resembled the body and legs of a crab to early medical worker Hippocrates in 400 B.C.

One of man's oldest stories of the crab relates to the constellation Cancer. A gigantic crab (named Cancer) was entrusted by Poseidon to guard the sea nymphs and through a series of events ended up in a fight with a giant squid. He became mortally wounded despite his immortal condition. To prevent Cancer's eternal suffering, Poseidon placed him among the stars. This constellation became the fourth sign of the zodiac five thousand years ago. The sign Cancer represented the love of home. Crabs have decorated coins since at least 425 B.C. and there are dozens of crab coins available at the click of a button to today's collector. Coins have been decorated with freshwater crabs, land crabs, hermits, and coconut crabs. Most coin designs relate to the animals, but many refer to the constellation Cancer.

Ancient Greece, 425 to 406 B.C.
Sicily, Akragas.

Humans have made countless toy crabs, nearly all of them are toys for toddlers. There was a popular Beanie Baby crab I picked up from McDonald's in 1999. Transformers toy crabs were also represented in the late 1990s. The character Razorclaw changes into a relatively life-like crab while one of the most famous Transformers Beast Wars characters (Rampage) is a transmetal crab. Large plastic crabs are used for decorations in beach-themed hotels, restaurants, and home décor, but they are not toys.

It is likely humans have kept crabs as pets, at least for a few hours at a time, as long as humans and crabs have shared the planet. However, keeping crabs as long-term pets in glass

Top (l-r):
1. Hermit Crab, Republic of Seychelles, one rupee, 1982
2. Red Crab (*Gecarcoidea natalis*), Christmas Island, fifty cents, 2016
3. Edible Crab (*Cancer pagurus*), Bailiwick of Guernsey, one penny, 1986

Bottom (l-r):
4. Crab, Fimmtiu Island, Iceland, fifty kronur, 1987
5. Coconut Crab (*Birgus latro*), Ripablik Blong, Vanautu, ten vatu, 1983
6. Maltese Freshwater Crab (*Potamon fluviatile*), Malta, five cents, 1998

boxes only became popular in the early to middle 1800s. Marine spider crabs and hermits are among the species with a long-written history as pets (Butler 1856, Mellen and Lanier 1935).

My earliest memories of crabs included handling fantastic creatures in the hands-on tide pool at Sea World of Ohio. There were a few different types, even horseshoe "crabs," but my favorite were the long-arm crabs. These were mottled tan, with a body a little smaller than a fiddler crab, but with spectacularly long, thickened, front arms terminating in harmless little pincers. Hermits and small fiddler crabs at the local pet shop and marine crabs were the next fascination. These were later replaced by the fantastic moon crabs encountered at reptile shows every so often. From 2006 to today a greater variety of crabs have shown up, including a few that can be reared in captivity without extreme efforts.

Razorclaw, Beast Wars Transformers® 1997

One of the most informative parts of keeping living creatures (rather than collecting dead curios such as seashells, vintage toys, or baseball cards) is being able to observe their habitat and behavior when you catch or collect them yourself. One can fondly remember the day an animal was found and take notes about the habitat it was living in, or record behavior that might never be seen in a captive enclosure. I have encountered a number of fantastic crabs during vacations along the east coast of the United States. Some

Silver bar with cancer logo

Covered urn, 1885-1900, decorated with signs of the zodiac, including Cancer

Various crabs can be found even on overpopulated commercial beaches. Myrtle Beach, South Carolina. © Sara Howard

Myrtle Beach tide pool: Mole Crabs, Hermits, and Speckled Swimming Crabs can be encountered within feet of each other. © Ryan McMonigle

An ancient pastime, *Bathers Playing with a Crab*, Pierre-Auguste Renoir circa 1897

are popular pets, while others are not or should not be kept in captivity. My fondest childhood memories of the beach were digging through the sand to capture *Eremita talpoida,* sand fleas (Atlantic sand or mole crabs), as the waves rushed out. Sand fleas are extremely difficult to keep alive in captivity for long periods. The speckled swimming crab (*Arenaeus cribrarius*) is a common though difficult creature to see or catch, but easy to maintain. It can be seen rapidly swimming sideways, just beneath the foamy waves in a few feet of water. White hermits (*Pagurus longicarpus*) may be encountered in sandy tide pools. Ghost crabs (*Ocypode quadrata*) may still be found on some popular beaches. Horseshoe crabs (*Limulus polyphemus*) are common in sand and mud flats, though are rarely seen on commercial beaches. *Panopeus herbstii* is found in muddy and rocky areas, underneath rocks at low tide. A trip to tropical beaches or forests can lead to far more fantastic finds. Of course, local laws, state licenses, or importation documents make taking pictures the most realistic method of catching animals or capturing their behaviors, especially if the same species can be found at home at the local pet shop.

King Crabs are rarely seen live at markets. This five-pound specimen was available for $33.99/lb. at an Asian market in Cleveland, Ohio. The photo is not out of focus, the glass is coated in condensation because the water is refrigerated.

One cause of serious problems with keeping any animal as a pet is that we often want to learn more about them. This invariably includes knowledge of the problems they face in nature, problems which are nearly always human in origin. Hobbyists then may voice concerns about habitat loss and destruction, at which point the response is always "keeping them as pets should be stopped"—even when the problem is known to be habitat loss, pollution, or commercial fishing. The pet trade is relatively harmless at its worst and is almost never a contributing factor, but it is the only factor that is easy to eradicate. This is the case even when available invertebrate stock is entirely captive-bred, such as with *Poecilotheria* tarantulas in U.S. holdings—the kickback from hobbyist concerns about habitat loss in India simply provokes efforts to make captive-bred specimens illegal in North America. It is too expensive and difficult to address habitat loss, even within one's own country. Consider the billions raised to save the rainforest, whose decline is unchecked. The unfortunate effect is when all the natural habitat is gone there may be no specimens in captivity and there will be no way to save lost species.

Conservation is often discussed with unrealistic altruism and limited honesty. Can we say humans save what they do not love and can humans love something that is unfamiliar? Is overcollection possible when there is adequate natural habitat for species that produce hundreds or millions of offspring in one go? Does collection make a difference if there is little to no habitat capable of sustaining a particular species? Is the science of sustainable harvesting for food remotely foolproof?

Overcollection is possible for large species, especially those humans like to eat. The terrestrial coconut crab (*Birgus latro*) was common in Australia and Madagascar before humans arrived on the scene. Humans love to eat crabs, but mostly the big ones. Management of harvesting is a good idea, but the exact number that can be removed before population collapse is difficult to implement and control. All the large marine crabs can be aqua-cultured with some effort, but there is limited economic incentive. A variety of species of freshwater crab are economically significant food items in Africa (Cumberlidge 1999) and Asia—*Somanniathelphusa* and other rice paddy crabs—and they have been consumed in quantity for centuries, if not millennia. Still, the human population is ever increasing and habitat ever decreasing.

Jonah Crab (*Cancer borealis*), ready to be boiled

Small crabs, even the most popular pets, do not face the same pressure as food species because they are not measured by the ton. A continuous demand for pet species may protect some habitat as collectors understand they lose their income if someone else tears up that habitat for industrial or farming purposes. When it comes to small crabs and immature specimens, it is very hard for humans to find them all—I have been unable to locate immature vampire crabs for six months in a mossy habitat not much wider across than my hand. Aquaculture could be employed for most terrarium species if demand were ever to make the costs warranted. Nevertheless, if aquaculture results in discontinued export, the former collectors must look for another way to earn a living and those endeavors might irreparably destroy habitat (and nobody would be the wiser). Even solutions bring their own problems. Nevertheless, increased knowledge about the husbandry of each species can be of perpetual value.

Crabs are not simply victims in the natural world. Their vast numbers, intense voraciousness, long history, and widespread prevalence means, as a group, they have likely molded many of the earth's ecosystems and extirpated more species than any other single type of creature. The tendency for crabs to eat other organisms and affect ecosystems is readily observed when humans accidentally introduce them to new areas. Habitat degradation can affect the way native species react with their environment. Indigenous purple marsh crabs (*Sesarma reticulatum*) are considered an important factor in the destruction of salt marshes across the eastern United States seaboard as a secondary effect of habitat degradation (Lewis 2007). Native *Ocypode* are considered the primary natural predator of loggerhead turtle nests in Australia (Trocini 2013).

Even the car-sized, giant crab monsters of Jules Verne's *Mysterious Island* (1961) were just being reared as food and of course they were reportedly quite delicious. Though there have been a few movies (*Attack of the Crab Monsters* 1957, *Island Claws* 1980, *Queen Crab* 2015) with giant crabs as villains, they just do not scare people any more than do vampire chickens or zombie cows. Cartoon characters over the years like Mr. Krabs, Sebastian, and Zoidberg are always cute and funny creatures. Though they rarely feature crabs, nature shows that include crab footage are legion. The recent *Bachelor in Paradise* shows included a lot of moon crab footage because

Coconut crabs are eaten throughout their range. Some populations have been eaten to extinction. The crab above is from Diego Garcia, British Indian Ocean Territory. (Drew Avery)

On Christmas Island, Australia, coconut crabs can be a traffic hazard. (*left*, David Stanley)

crabs symbolize life on a tropical beach. In coastal areas and Caribbean-themed hotels, plastic crabs often decorate the walls and likenesses are knitted on decorative patio pillows. There are giant crab statues in San Francisco, California, Krabi, Indonesia, and Kep, Cambodia, of uncertain origins, but they certainly all pay homage to the economic food value. A very large blue crab statue hangs above the massive entryway of the Norfolk Aquarium in Virginia in tribute to the fishery value of this species.

Crab snacks—even small species can be fried and spiced.

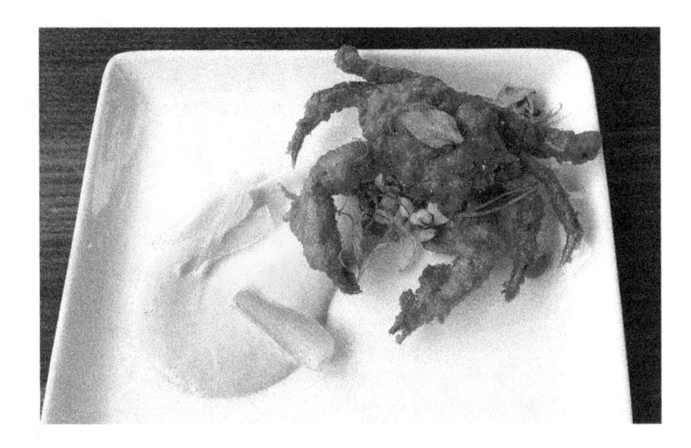

Fried soft-shelled crab (T. Tseng)

This is the only pet I also love to eat. I have had my heart set on crab cakes served in a flipped-over blue crab carapace since childhood, and yet it does seem a little wrong when keeping the same animal as a pet. Crabs are significant food items across the world. Snow crab (*Chionoecetes* spp.), blue crab (*Callinectes sapidus*), Dungeness (*Metacarcinus magister*), and the anomuran king crab (*Paralithodes* spp.) can be found in most North American grocery stores. Soft-shelled (freshly molted) crabs can be eaten whole and are a major delicacy. Asian markets (those in North America) offer some of the larger food crabs live and carry others not seen at the standard grocery store, like rock crabs (*Cancer* sp.), deep-sea red crabs (*Chaceon quinquidens*), and crystal crabs (*Chaceaon bicolor*)—the latter was believed to be the same as the species found in West Australia (*C. albus*) until 2007. Markets in Europe generally carry a few large species, while smaller species and terrestrial crabs like coconut crabs, *Cardisoma* land crabs, and rice paddy crabs are available at markets across Asia.

Steamed crabs at a fish market (Alan Kotok)

Eating crabs is not just about the food value. Catching crabs is a common pastime along the coast, even though the crab pots, raw chicken bait, and pier charges usually rack up more costs than you would pay at a seafood counter. The name most people who catch crabs give themselves, is *crabbers*. The name for people who keep crabs as pets is the same.

Kep, Cambodia
© Maurizio Biso

Krabi, Thailand
© Anurak Anachai

Roadside monument to the land crabs that cross the Varadero-Cárdenas highway in Cuba (Kurt Bauschardt)

Temple carving in Thailand © Wetchawut Masathianwong

New Crab, Klawock Totem Park, Alaska (Mack Lundy)

PHYLOGENY, MORPHOLOGY, AND BIOLOGY

Phylum: Arthropoda = Arachnids, crustaceans, insects and other joint-legged creatures.

Subphylum: Crustacea = Barnacles, ostracods, triops, malacostracans, and others with branched limbs and planktonic larvae.

Class: Malacostraca = Isopods, mantis shrimp, decapods, and others with three major body parts (tagmata).

Order: Decapoda = Hermit, king, porcelain, and true crabs; true shrimp, lobsters, crayfish, and other ten-legged malacostracans.

Infraorder: Brachyura = True crabs, nearly 7,000 species with greatly reduced abdomens and a squat body shape.

Family: Sesarmidae = 29 genera of mostly terrestrial crabs, some that do not need to return to the sea for spawning.

Genus: *Geosesarma* = 53+ species of small, often colorful, true terrestrial crabs with truncate larval development.

Infraorder: Anomura = Hermit crabs, king crabs, porcelain crabs, mole crabs, and squat lobsters.

Family: Coenobitidae = 2 genera: *Birgus* and *Coenobita*

Genus: *Coenobita* = 16 species of terrestrial hermits.

The red devil (*Geosesarma hagen*) is one of the most popular vampire crab species today.

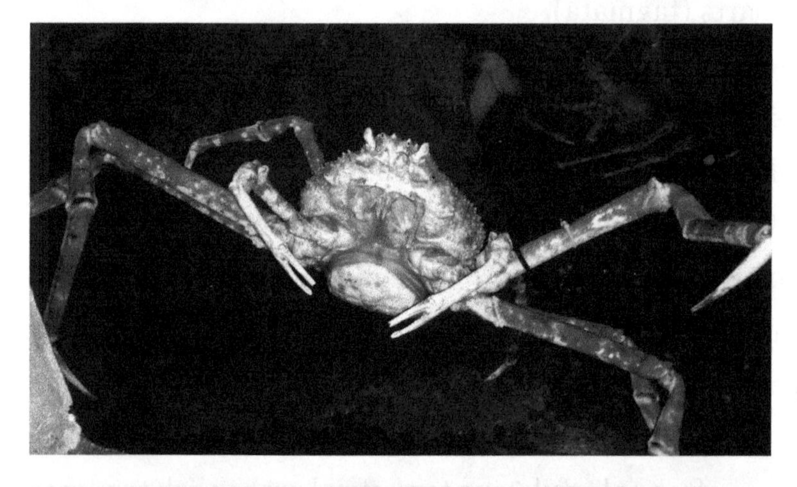

Giant Japanese spider crab (*Macrocheira kaempferi*). Though the legspan of this female is only a little over a meter, the males have massive, elongate arms with a span that can reach eighteen feet (5.5 m).

Terrestrial hermit crab, *Coenobita cavipes* © Christian Schwarz

The mole crab (*Emerita talpoida*) is an unusual anomuran crab that can be found buried in beach sand.

Porcelain crabs like this *Petrolisthes* sp. are anomuran crabs like hermits and mole crabs.

Uca minax male fiddler crab

Cherry shrimp (*Neocaridina davidi*) is a small freshwater shrimp that is easily bred. This species started the micro shrimp craze.

The sea slater (*Ligia oceanica*) is a malacostracan, but it is an isopod, not a decopod like crabs and shrimp.

Slipper lobsters are decapods seen at marine shops from time to time; like true lobsters they only produce planktonic larvae.

Red claw shrimp (*Macrobrachium assamense*) is the most commonly maintained large shrimp. Unlike many members of the same genus it has large eggs that hatch into small shrimp rather than planktonic larvae.

Crayfish have gained some popularity as pets in recent years. *Procambarus* sp., Florida.

Triops cancriformis, a branchiopodan crustacean often reared as a novelty pet. The dried eggs are sold in envelopes. The time span between hatching and death can be less than two months. *Triops* superficially resemble horseshoe crabs, but are related to fairy shrimp and brine shrimp.

Relatives of the Brachyura (true crabs) include all of the ten-legged crustaceans in the decapod group and nearly all the familiar marine arthropods, including shrimp and lobster. Another decapod infraorder, the Anomura, also has many members commonly known as crabs. This group includes the porcelain crabs, king crabs, mole crabs, squat lobsters and hermit crabs. Nearly all the other so-called crabs, like crab spiders, crab lice, and horseshoe crabs, are not crustaceans and so are distantly related. Not to be confused with the crab spiders (arachnids), the members of certain true crab families are commonly known as spider crabs.

The biggest and most fantastic arthropod living in the earth's oceans is the Japanese spider crab. Males can measure 12 ft. (4 m) from claw to claw. Earth's largest terrestrial arthropod, a close relative of the common pet hermit, is the 3 ft. (1 m) legspan coconut crab. On the other end of the spectrum, tiny marine pea crabs live as parasites inside clams and mussels and are sometimes considered the smallest crabs (legspan as little as 10 mm). The smallest adult pea crabs are larger bodied than a normal freshwater *Limnopilos* crab, but the legs are shorter. The smallest crabs are still massive compared to the smallest beetles, wasps, flies, moths, or springtails.

Morphology

Arthropod exoskeletons are made of chitin (a substance similar to cellulose) along with some protein and mineral content. Chitin is a versatile material formed into sensory setae (hairs), spines, ocular lenses, and amazing body shapes and appendages. It can form pliable, thin membranes which hold moving body sections together —or, it can be thickened, reinforced with other materials, and formed into rock-hard plates or tubes. Crabs have some of the thickest and hardest shells of all the crustaceans.

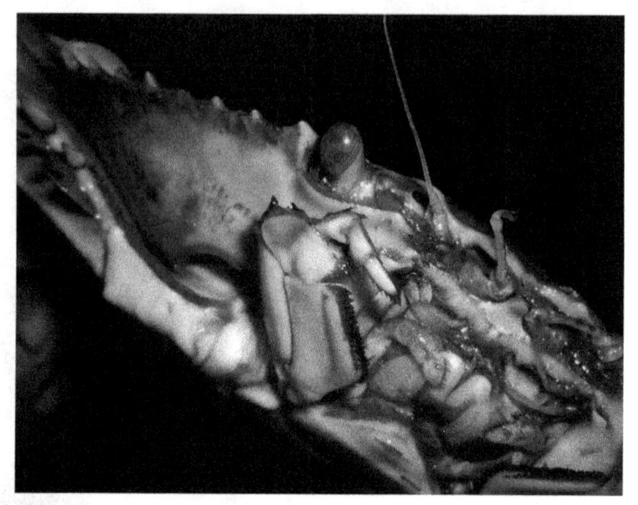

Compound eye, antenna, and antennule of a blue crab

The hardened exoskeleton is primarily impregnated with calcium salts. The mere presence of calcium does not make the exoskeleton hard since the softshell (teneral) crab has the same mineral content within its body before and after the shell hardens. The carapace of *Cancer pagurus*, one of the thickest and hardest shelled of all crabs, is comprised of 22% calcium and 1.2% magnesium by weight (Boßelmann et al. 2007). Other species may have considerably less mineral content. Discarded exoskeletons burn at about the same temperature wood burns and become a tiny pile of black ash that is a fraction of the original mass. Marine species extract calcium from seawater. Freshwater species can extract calcium from water, but fresh water does not always have available calcium. Terrestrial species acquire calcium primarily from their food. Terrestrial and freshwater species usually eat their old shells following the molt and will chew up snail and bivalve shells.

The crustacean body is divided into two or three main sections depending on whether or not the head is fused to the thorax. In true crabs, the fusing of the head and thorax is so complete it is not possible to see where one begins and the

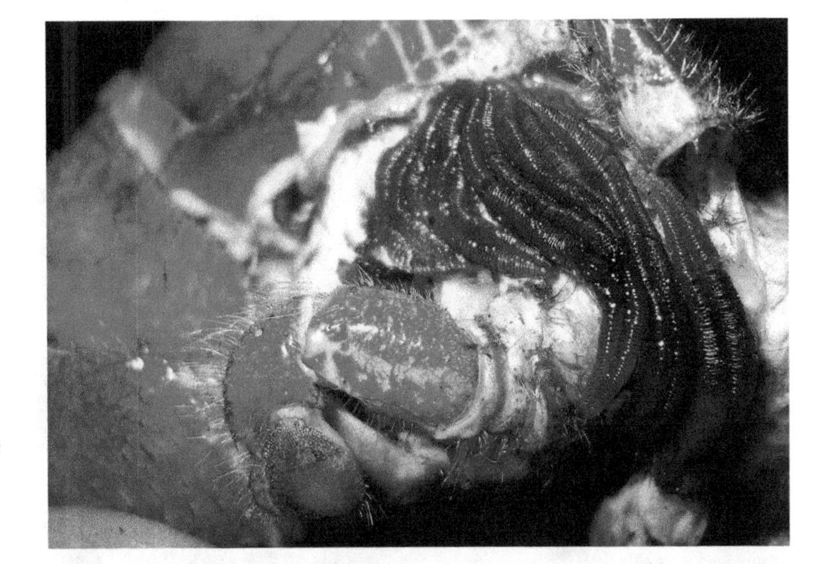

Hermits, including this terrestrial *Coenobita*, have large gills located under the carapace. Like true crabs, the gills originate from near the base of the legs.

Large gills are located under the carapace near the bases of the legs. These gills are part of a moon crab exuvium and are shed at each molt.

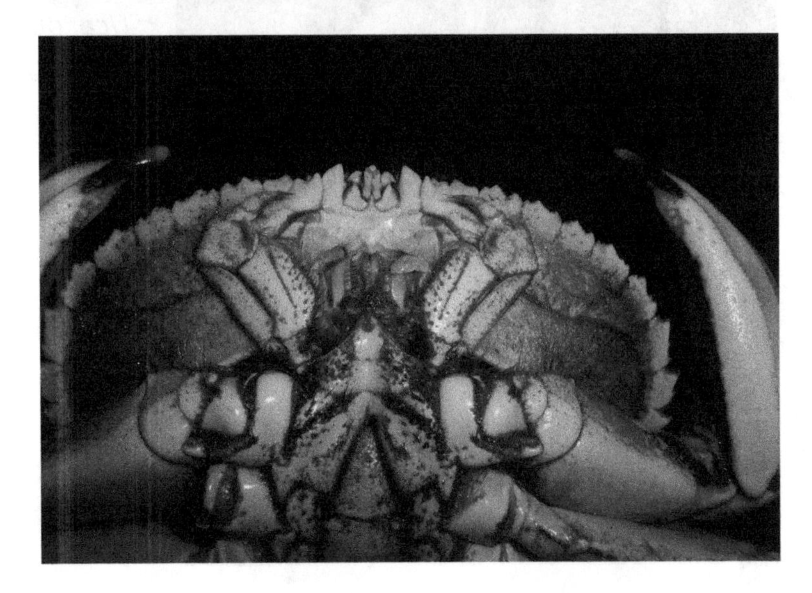

Mouthparts: maxillipeds, maxillae, and mandibles are visible on this *Cancer borealis*.

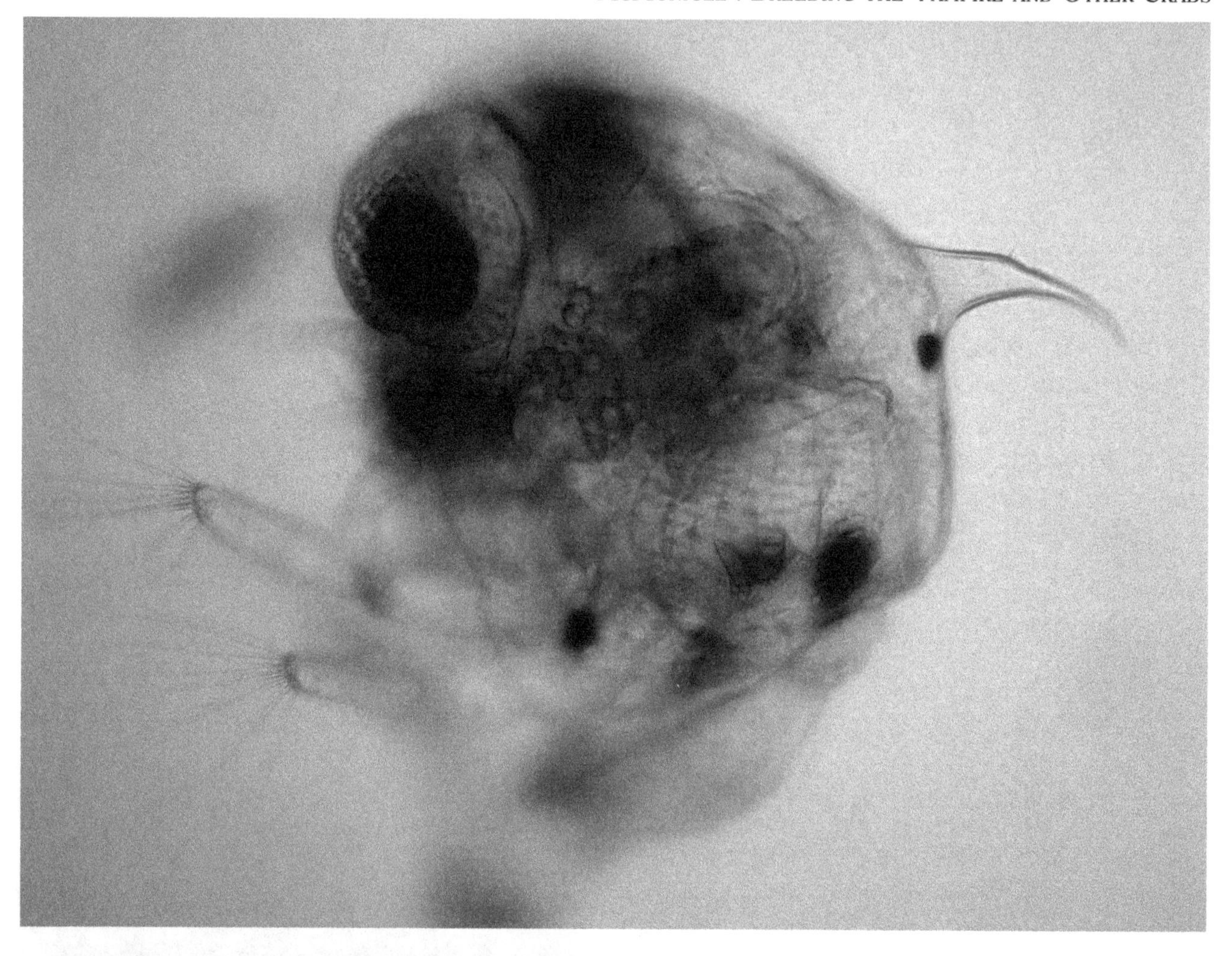

The compound eye is comprised of hundreds or thousands of ommatidia. Even in early development (2nd instar), this *Uca pugilator* has hundreds of ommatidia making up each eye.

The rear legs of a hermit crab can be seen here.

other ends. This main body section is known as the cephalothorax and it contains the antennae, eyes, mouthparts, legs, internal genitalia, digestive system, and gills. The tail (abdomen or pleon) is a small section folded underneath which holds only the genitalia, pleopods (abdominal appendages used chiefly to hold onto eggs or mates), and the end of the digestive system. The abdomen of hermit crabs has the same basic structures, but it is usually soft, more massive, and protected by the shell of a dead snail.

The outer pair of antennae are similar to those of insects, but crustaceans have an additional pair of antennae. The inner pair, the antennules, are usually branched and often well-articulated for multidirectional movement. The central antennae tend to move in rapid flickering waves and can fold up into a recess in the carapace. The presence of two pair of antennae is an easy way to tell a crustacean from other types of arthropods, but terrestrial true crabs (like vampire crabs) and many of the semi-aquatics (like *Uca* spp.) have reduced antennae that are very difficult to see. Anomurans (hermit crabs, porcelain crabs, and squat lobsters), whether they live underwater or on land, usually have very long, pronounced antennae and shorter antennules.

Crustaceans are known for the most extreme development, largest numbers of facets (ommatidia), and the ability to see more colors than other creatures. The compound eyes of crustacea are seldom reduced and all of the commonly maintained crabs have excellent eyesight. They

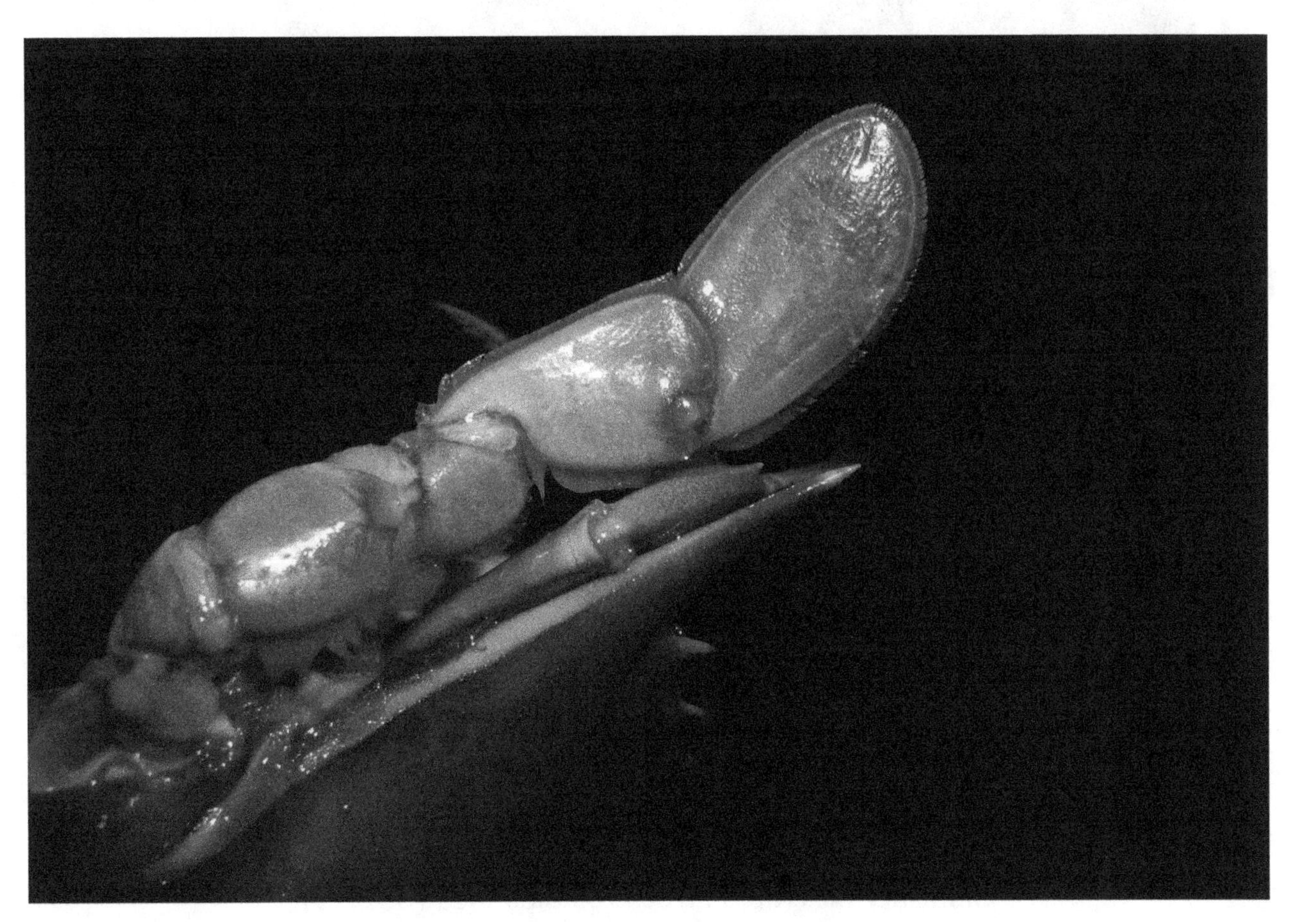

The paddle-shaped rear leg of a blue crab

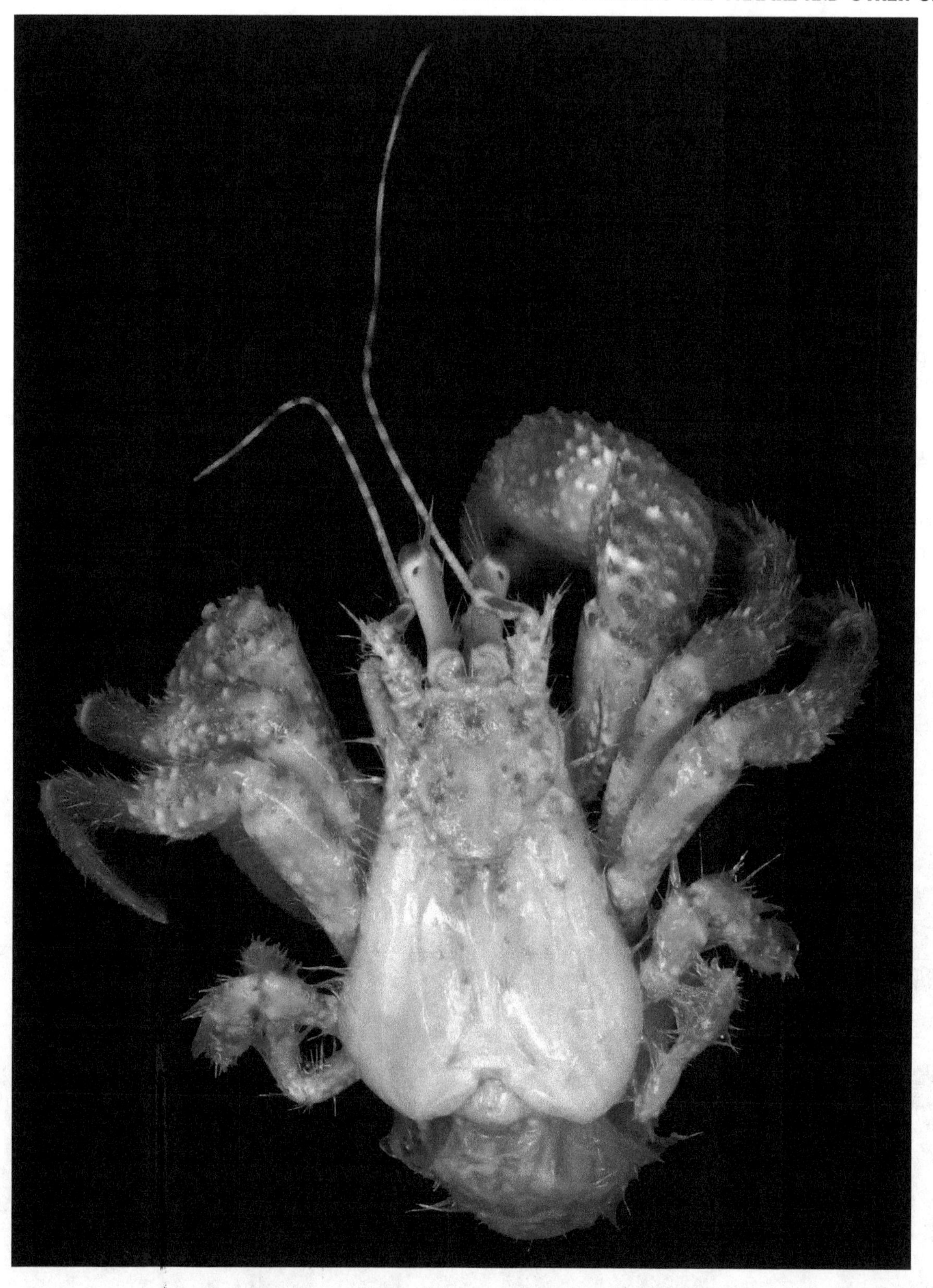

Giant hermit crab, *Petrochirus diogenes* (USFWS)
(*above*) dorsal, (*opposite*) ventral

use vision as a key tool for survival: to avoid predators, to find food, and to identify healthy mates. Most species notice movement from far away and react to their keeper entering the room. They may shuffle rapidly into hiding, freeze in place, or slowly sink back into their burrow. The eyes are often used to identify the movement of prey as well as predator. Some are attracted to rapid movement of tiny objects, like a house cat is. Unlike insects and arachnids, crustaceans normally only have one pair of compound eyes without any large accessory simple eyes (ocelli). While most arthropods have fused, stationary eyes, the eyes of crabs are often located on stalks articulated for movement. Stalked eyes can fold into a recess in the exoskeleton for protection. If damaged or removed, the eyes may be able to regenerate like the legs.

Maxillipeds, the largest, outer pair of mouthparts, and two inner pairs, are adapted for feeding and transporting water to the gills. Filter-feeding species can have massive, hairy appendages used to strain tiny food from the water (or to scrape food from feathery antennae). In true crabs the maxillipeds circulate water to the gills. Maxillipeds arise from the trunk rather than the head, so they are legs adapted as mouthparts, but they are not called legs.

Crabs are decapods (*deca* = "10" and *podus* = "foot") and have five pairs of thoracic legs. The first pair of legs (the front arms or claws) are known as chelipeds. These often terminate in massive pincers (chelae) or consist of large, elongate basal segments with small pincers at the end. Pairs two through five are usually called the walking legs and none end in pincers like they do in shrimp and crayfish. The third pair (including the chelipeds as pair one) can have indentations or holes near the base where the female accepts sperm. The fourth and fifth pairs are sometimes very small and used for purposes other than walking. The last two pairs of legs of hermit crabs are small and used to hold the shell in place. Sponge crabs (Dromiidae) have larger rear legs adapted to hold a hollowed-out sponge over their carapace. In porcelain and king crabs, only the rear legs (fifth pair) are reduced and hidden within the carapace. In swimming crabs, the rear pair is flattened into paddles. In a few crabs, all four pairs of walking legs are paddle-shaped for swimming. The first or fifth pair sometimes has ridges used in stridulation. The chelipeds and walking legs together are known as pereopods because they are attached to the thorax (*pereon* = "thorax" and *podus* = "foot").

The additional leg-like appendages on the abdomen are called pleopods (*pleon* = "abdomen" and *podus* = "foot"). These are mostly used in reproduction with feathery appendages on the female used to hold eggs and leg-like appendages on the male used during mating. Unlike lobster, shrimp, mantis shrimp, and other crustaceans, crabs do not use the pleopods for swimming or walking (except during the megalopa planktonic stage). Instead, the fifth pair of thoracic legs (pereopods) are adapted into paddles for swimming crabs.

The Chameleons

Though crabs were almost certainly changing color before the first ancestor of the first chameleon, the ability to change coloration to match the background is known best in the famous lizards. Many different true crabs, especially shore crabs like fiddler, marsh, and ghost crabs, can darken or lighten their body color to match the background. Color changes usually require a few hours and do not occur in seconds as with chameleons or cuttlefish. Mud crabs and others also change color, but a molt is required for a significant color change to match the background contrast. Blending into the background likely helps crabs avoid being eaten by birds, sea turtles, otters, and other crabs. Immature crabs tend to

Color changes can be slow and take weeks or require a molt for the animal to adapt to the background colors. The mud-colored crab on the left will not adapt to the new background until it molts. Dwarf mud crab (*Rhithropanopeus harrisii*).

change color more readily than adults, especially swimming crabs and terrestrial freshwater species. *Geosesarma* immatures are very good at blending into the background, while the adults are brightly colored and have little or no ability to change.

Not all color changes are related to crypsis. The adult *Metasesarma aubryi* (called the red chameleon crab in Germany, but known as the red apple crab in the U.S.) can have very different color patterns, from plain dark brown to ridges of orange splashes or even solid red. Daytime changes are in response to the background color and lighting, but the most substantial changes take place at night. Other sesarmids like *Armases* and *Geosesarma* (immatures primarily) change colors between night and day. For most crabs, the shed exoskeleton (exuvium) is mostly white or orange. As the new and old exoskeletons begin to separate, crabs transform to a pale or brownish hue a few days before the molt. One of the most familiar color changes is not related to the environment or to growth; nearly every crab turns red when cooked. This change occurs because the dark and green pigments cannot stand up to extreme heat (not just so we know when they are ready to be eaten).

Although most brachyurans can change color for at least part of their lives, it is not a common feature of anomuran crabs. I have never seen a hermit or porcelain crab change color unrelated to a molt. Terrestrial hermit crabs can be pale

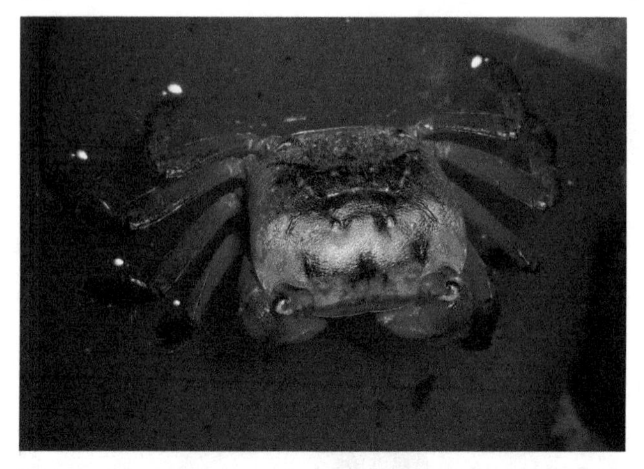

The red apple crab, *Metasesarma aubryi*, female is usually more variable in coloration than the male.

Red apple crab, same female: lighter colors are usually seen in the morning.

Red apple crabs: on this day the female (on right) looks very much like the male.

Red apple crabs: the female (on left) is the same animal on a different day.

when small and darken as they grow. A significant change in diet can lead to a change in color at the next molt.

Behavior

Crabs exhibit many behaviors to help them avoid predators and other dangers. Aquatic and terrestrial species dig burrows to protect themselves from daytime predators, changes in the water level and turbulence, or the drying sun. Terrestrial crabs may dig a few feet down to reach water if there is none within walking distance. Some do not have burrows, but bury themselves in sand or mud with incredible speed. Hermit crabs carry their shelters around with them for a hasty retreat but still can construct burrows. For terrestrial species desiccation may be the most important driver of nocturnal behavior. *Geosesarma* and other terrestrial crabs are very active during and after rains in their natural habitat. Terrestrial and semi-aquatic species are usually active from dusk till dawn when their behaviors are less likely to be seen by day-active land predators such as birds and mammals. Aquatic species are also night active to escape sea turtles, fish, and octopus. Some species play dead when handled.

Others move rapidly or use their powerful claws and aggressive attitude to protect themselves. They may exhibit a defensive posture, display contrasting colors, and raise sharp, open claws to the attacking predator. Most crabs have excellent eyesight and run away and hide when the keeper (or predator) approaches.

Crabs often live and interact in groups. They communicate with each other through sound and sight. Nearly all the true crabs have excellent vision and are able to detect color and movement with precision. The prevalence of sound in intraspecific communication (such as the chirping of *Coenobita compressus*, drumming of *Uca pugilator*, and squeaking of *Sommaniathelphusa* sp.) seems to be relatively rare, though sound could be used as much in water as in air and go unnoticed by humans. Living in groups aids in finding mates; it is also a useful strategy in predator avoidance. Two of the most common pets, hermits and fiddlers, are known for their gregarious habits in nature, but they do kill each other in captivity under certain conditions.

There are a number of different levels of cohabitative behavior. The lowest level species tend to kill each other when placed in the same cage. The highest level never cause any harm to each other. Sometimes the ability to live together relates to dietary preference, enclosure properties, gender, molting, or individual behavior; aggression is usually similar for members of the same species under similar conditions. The effects of common species cohabitation is arbitrarily listed below in order from most aggressive (0) to least (7). An explanation for the rating is provided.

0.) It is usually possible to put two patriot crabs together for a few minutes before they begin to tear off each other's legs. Ghost crabs are not nearly so quick to kill if given space, but they can attack and harm each other. Adult male red claw crabs usually do not damage other adult males. However, adult females of the same species (and other crab species placed with them) will have their legs torn from their bodies.

1.) Swimming crabs do not usually bother each other once the exoskeleton has hardened. Immediately following the molt, they are susceptible to injury and death. Appendages or the entire bodies of recently molted tank-mates are eaten. Small crabs of the same or unrelated species can be consumed whether or not they have molted.

2.) Panther crabs can live together safely in an aquarium if they are provided space and individual hides, but they do not like to be within arm's-length of each other. Fighting may occur and result in the loss of a front claw. Freshly molted specimens can be eaten but I have not experienced such an event. Dwarf mud crabs often hang out close to each other and molt close to

Hermit crabs famously gain protection from discarded snail shells.

Many crabs like this *Cardisoma* lift their claws in a threatening pose to ward off attackers.

Threat pose of the male *Armases cinereum*

Fiddler crabs raise their one big claw in defense if a predator gets too close.

From above, both male and female *Uca minax* blend into a brown background.

Some crabs are arboreal and escape predators by climbing up or jumping off into water or mud beneath the trees. Mangrove tree crab, *Aratus pisonii* (Melissa McMasters)

other specimens without being eaten. However, sometimes large specimens tear half the legs off of significantly smaller specimens immediately after the smaller crab molts.

3.) Vampire crabs get along as immatures, hang out together in small cages, and are usually safe to molt near specimens that are not greatly different in size. As they approach maturity they may fight violently and males may tear off each other's legs. Females are normally safe. Adults eat their own young (nearly every crab cannibalizes extremely tiny members of its own or other species that fit between the two fingers of a claw).

4.) Terrestrial hermits often hang out together. Even the males are not known to damage each other. However, large animals can cause small specimens to starve. Terrestrial hermits will kill any freshly molted animal that has not had time for the exoskeleton to harden. Red apple crabs seem gregarious, but they kill or maim each other during molting if there is not enough space (damage seems unintentional and they do not normally eat the other crab or its body parts). Squareback marsh crabs can be kept in groups of similar-sized individuals with little to no aggression.

This *Sundathelphusa* sp. has sacrificed one of its claws to escape. © Christian Schwarz

This female red fiddler plays dead when handled.

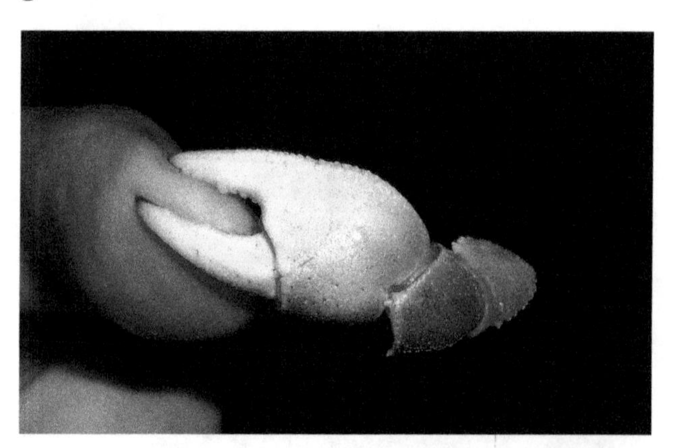

If the threat is unheaded and the crab must resort to pinching, the claw may detach as the animal runs away. This *Cardisoma armatum* cheliped retained its ability to pinch for several minutes after autotomy.

Red fiddler, normal stance

5.) Fiddler crabs almost never eat or harm each other during molts. Immature specimens and females do not damage each other in fights. However, mature males fight for dominance and the biggest male may eventually kill all the others, depending partly on the number of mature males and partly on the available space and hiding spots. At least some mangrove crabs (*Perisesarma*) live and molt safely in groups of males and females.

6.) The commonly available small marine hermit crab species fight over shells and food, but they do not kill other specimens, even during molts. Many different species can be kept together safely. This does not include large species, especially not very large types like *Dardanus* spp. that eat smaller hermits.

7.) Freshwater Thai microcrabs often cling to each other without causing harm. Marine emerald crabs and Sally Lightfoot crabs live in groups safely and molt near each other without harm. Males do not kill each other and relatively small specimens are not eaten by large ones.

Male and female dimorphism in a small freshwater crab, *Lepidothelphusa cognetti*

Reproduction

Most true crabs exhibit strong sexual dimorphism. Usually, the most obvious masculine feature is the enlarged front legs (chelipeds) used to battle over mates or to win mates through visual display. Extreme cases of cheliped dimorphism are seen in the major male signal crabs of most *Uca* species and *Lepidothelphusa* species. One claw (chela) can be as almost as massive as the other legs and body combined. In certain spider crabs (*Macrocheira, Mithrax*, et al.) the male's front legs can be many times longer and thicker than the female's. Minor males of strongly dimorphic species have less pronounced, but obviously masculine, claws. (Even in strongly dimorphic species the occasional female ends up with an enlarged or elongate cheliped.) Unlike giant stag beetles and some other insects where the adult male's weaponry is the ultimate expression of maturity, in crustaceans the minor males can eventually develop into major males. Even when visible differences between genders are relatively small, male crabs may have larger or differently shaped chelipeds that easily differentiate them from mature females (*Aratus, Armases, Gecarcinus, Geosesarma, Metasesarma, Percnon, Perisesarma, Pseudosesarma*, et al.). The other legs of the male may also be longer in relation to the body. Males can be much larger or smaller in overall size. The front or rear legs of some males have stridulatory apparatus, but these are just rows of tiny bumps that are not very noticeable.

When the chelipeds are similar in size, the female's pleon (tail) is usually an easy indicator of gender. It is almost always very wide and covers most of the belly (ventral surface of the carapace) while the male's pleon covers barely half or a quarter of the space between the legs. Beneath the tail, females have half a dozen or so large feathery pleopods. Males have one enlarged, bony pair of pleopods while the additional pairs are reduced or absent. Adult *Parathelphusa, Rhithropanopeus,* and *Syntripsa* females have narrow abdomens compared to other adult

Percnon gibbesi molts from both male and female show the differences in the abdomen and front claws.

female crabs, but they are visibly wider than the abdomen of a mature male of their own species. The mature male *Aratus pisonii* has a wide abdomen, but the telson is very narrow. Porcelain crab genders look similar from above, but like true crabs the male has a narrow abdomen while the female's is widened. Female porcelain crabs likewise have feathery pleopods to hold eggs beneath the wide abdomen. Immature specimens of all species have less pronounced or entirely lacking dimorphic features.

Hermit crabs show very little dimorphism at maturity and it is difficult to check for gonopores or pleopods because they are usually covered by a shell. The pleopods are on the abdomen while gonopores are seen as small holes in the basal segment of the third pair of legs. Gonopores on true crabs are located on the same segment of the third leg pair, which are visible on some species (the abdomen must be held open to see this).

Sexual dimorphism is not limited to physical structures of the exoskeleton; coloration and chemical signals are different. Females are often dull or cryptic in appearance, while males of the same species can be brightly colored, especially for true crabs that live along the shore or on land. This suggests the male's color and the female crab's keen eyesight play an important role in mate selection above the water. In marine environments the females tend to lure males through pheromones broadcast in the water. Few pheromones have been chemically isolated, though some have been identified in female blue crab urine (Kamio 2010).

Once they have located each other, the pair may stay together for some days or weeks, or begin the process of fertilization immediately. The duration and courtship of pairing varies widely by species and group. In many swimming crabs (Portunidae) the male rides around on the female's back for weeks, but it is not the act of mating—this behavior serves to guard her from other males. *Geosesarma* species often spend a week or two in the same tunnel. This could be before mating or between multiple attempts. The *Metasesarma aubryi* male holds onto the female's walking legs with his claws for many hours prior to mating. During mating, true crabs open up their abdomens, face belly to belly, and the male's gonopods direct the spermatophore towards the female's gonopores. Females of some terrestrial crabs blow bubbles during mating, which might prevent water loss of the spermatophore during transfer (McMonigle 2016). Hermit crabs also mate facing each other, but this is not always easy to distinguish from fighting and they do not completely leave their shells.

Observations of decapod mating are uncommon in captivity. This is partly because the average keeper's observations are almost entirely

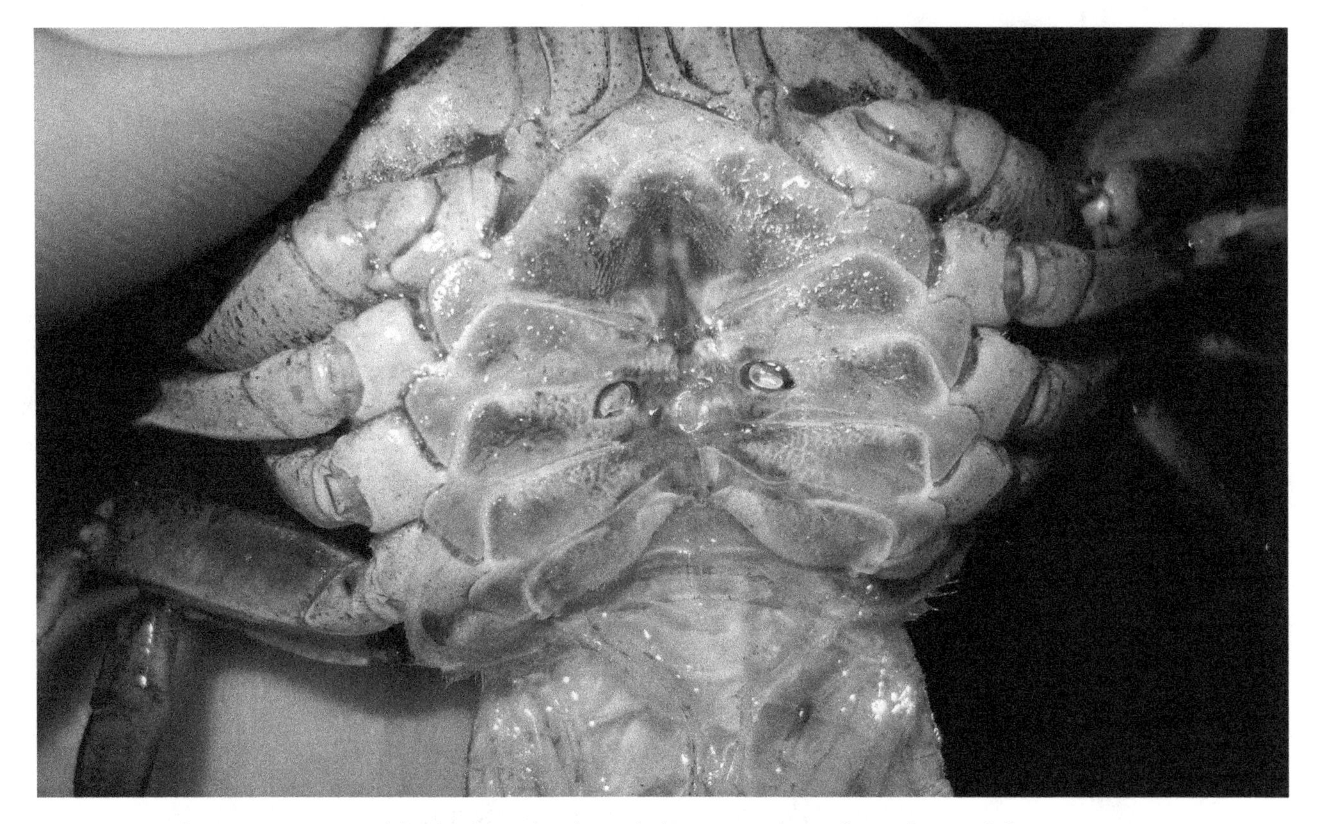

Gonopores visible beneath the abdomen, on a female crab's exuvium

Mithraculus sculptus, female abdomen (pleon)

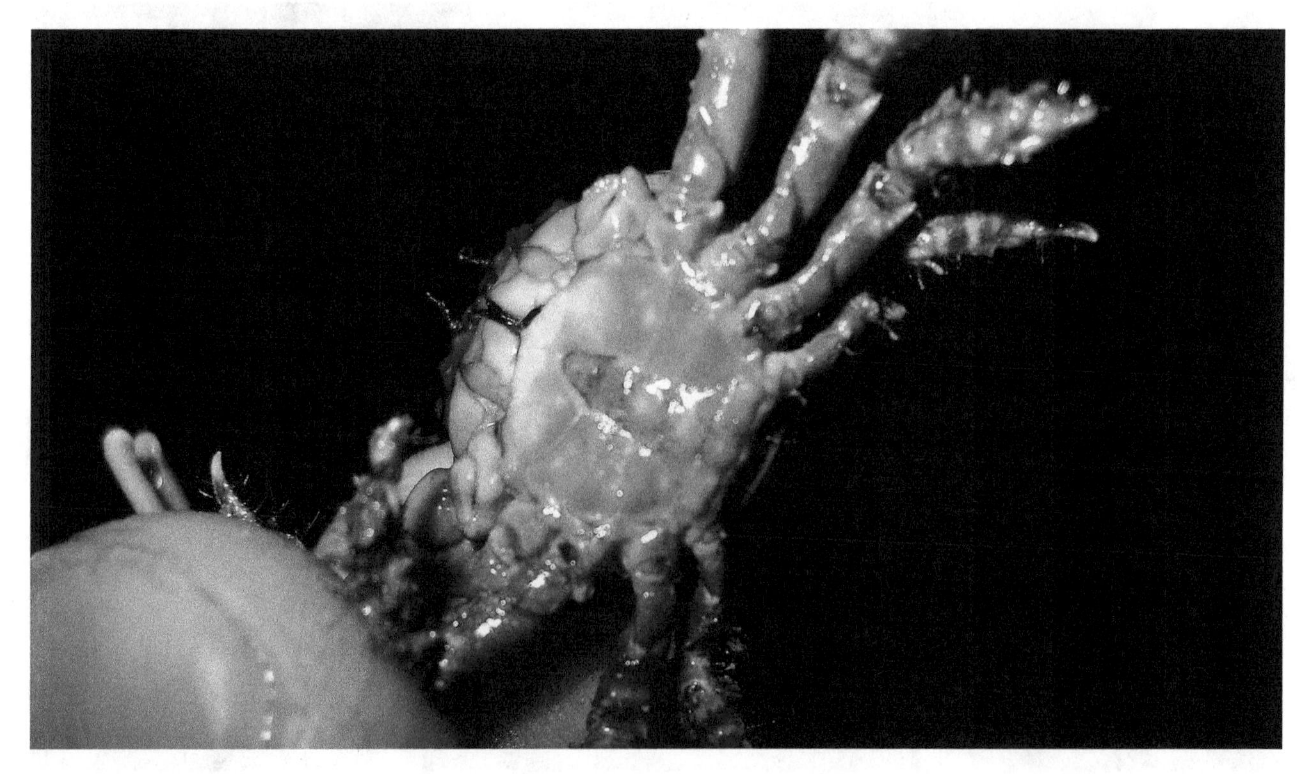

Mithraculus sculptus, male abdomen

Moon crab, female abdomen

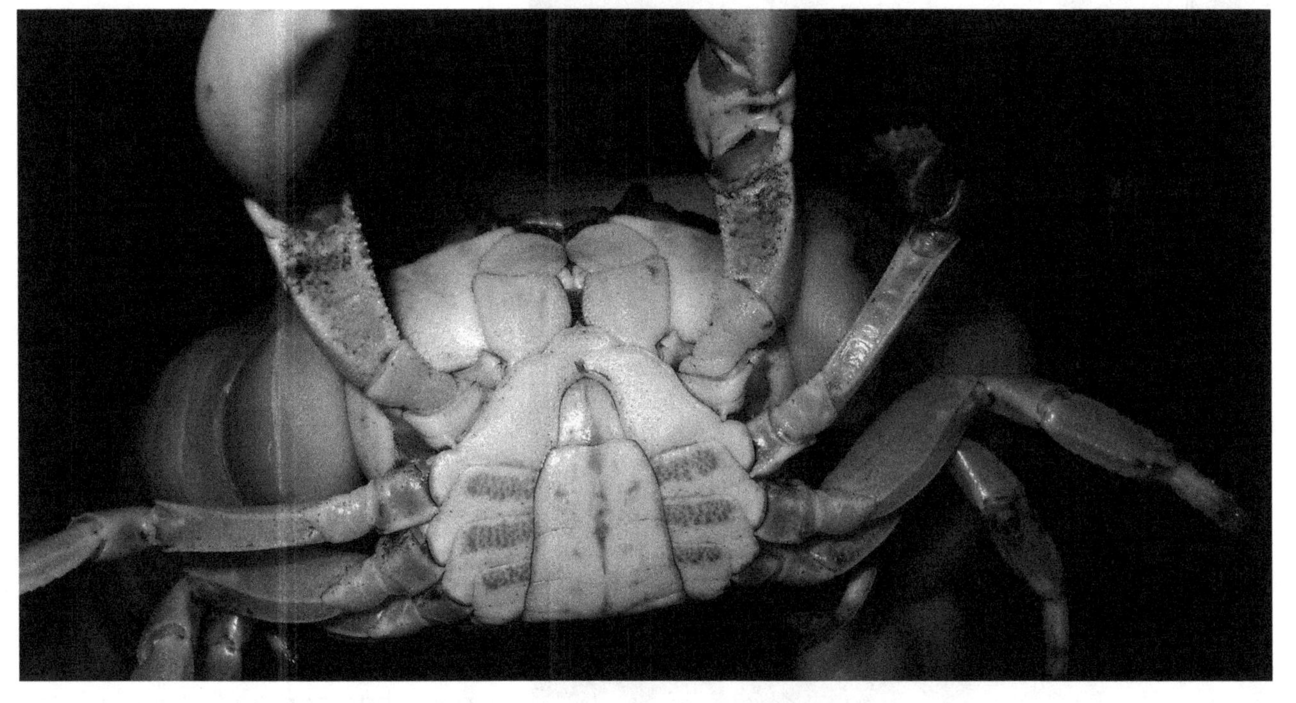

Moon crab, male abdomen

Uca puglilator female with a wide abdomen and normal claws

Uca pugilator female with an unusual big claw (appeared to be male from above)

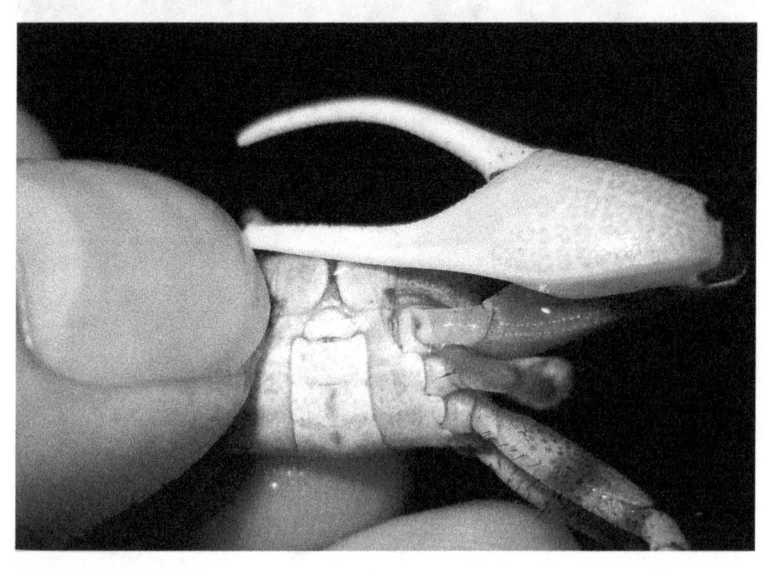

Uca pugilator male with a narrow abdomen and standard enlarged claw

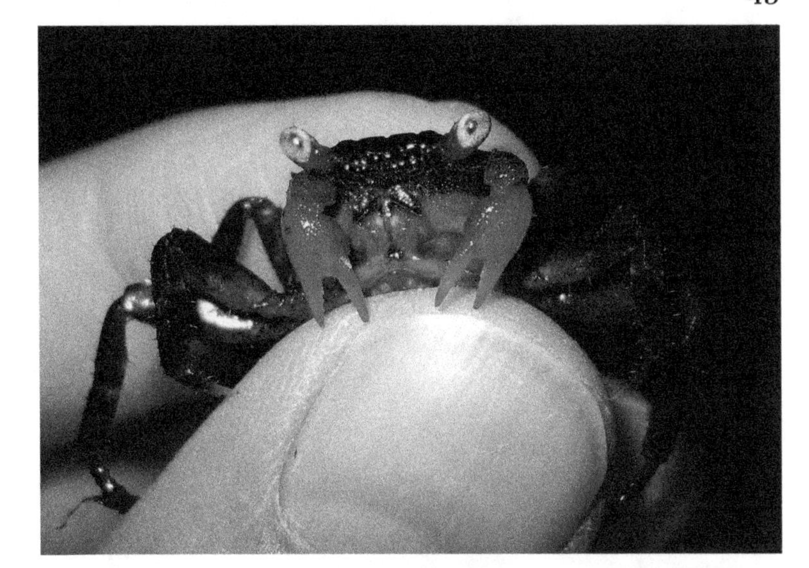

Red devil female with small claws

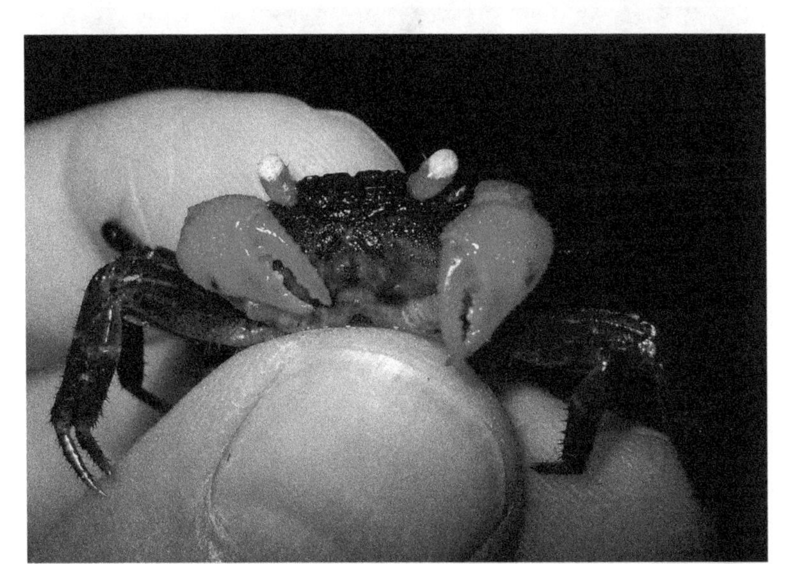

Red devil male with larger claws

The terminal abdominal segment of a female arrow crab is enlarged into a hardened case.

The adult orchid vampire male has larger claws.

Orchid vampire female with smaller claws and a chamber to gestate eggs

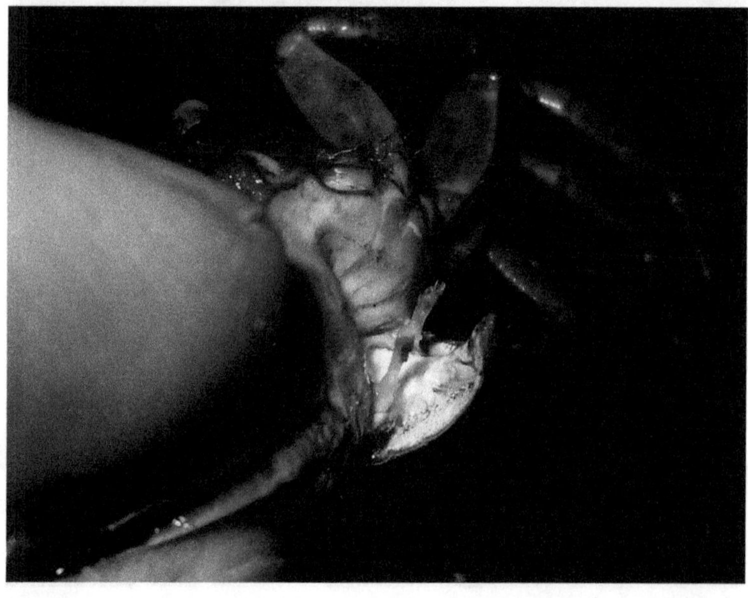

Gonopods of a male *Geosesarma*

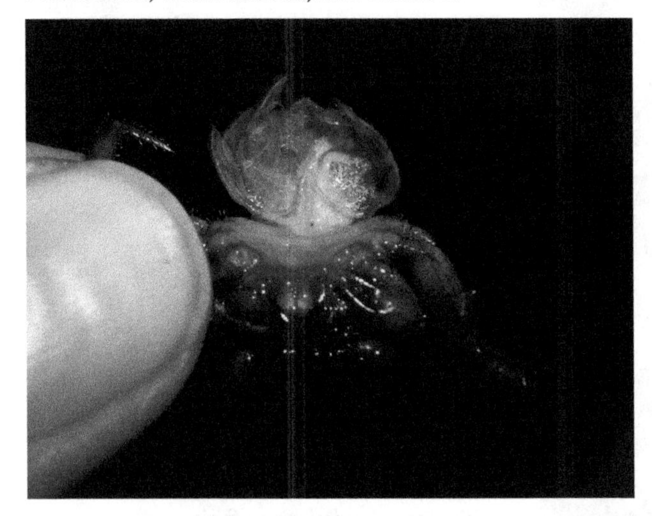

The female (an orchid vampire here) often has large, feathery pleopods to hold the eggs.

made when the lights are on—mating usually occurs in the dark. Another reason is that mating for most crabs happens just once or twice a year. I came across *Geosesarma* vampire crabs mating for the first time in 2016. (I had first mated captive-reared specimens in 2011, but had never witnessed the act of mating.) I was surprised by the foam production and the timing. I had always been told that shrimp, crayfish, crabs, and other crustaceans needed to be teneral for mating and neither male nor female had molted in months. I have raised *Macrobrachium* and *Neocardia* shrimp for more than a decade, but never noticed pairing. I have only witnessed one crayfish species, *Procambarus clarkii,* mating and the female was certainly not teneral. For true crabs I have also observed mating pairs of *Gecarcinus*, *Metasesarma, Mithraculus, Percnon*, and *Uca* (both *U. minax* and *U. pugilator*), none of which were related to molting. I have observed hermit crab mating and it was also not related to a recent molt. For these species a recent molt was not required and a recently molted female is likely unable to mate even if she could survive. Still, teneral mating is not a myth, some of the more commonly studied crabs (specifically the swimming crabs in the Family Portunidae) are known to mate only immediately after the female molts.

If one is familiar with the extreme sensitivity of teneral arthropods and the mating embrace of crabs and crayfish, it is difficult to imagine a mating female could survive teneral mating without her chelipeds smashed to uselessness

Male *Geosesarma* usually have roughly equal claws, the smaller claw on this male is the result of autotomy and regeneration.

Female *Geosesarma* have relatively smaller claws than males.

For comparison, mating *Procambarus clarkii* crayfish

In these mating sandfire vampire crabs, only the large chelae and gonopods of the male are visible directly below the female because he is in a tunnel. The foam seems to have been created by the female.

Mating *Uca pugilator* getting into position

Two male *Geosesarma* sp. in a Panay rainforest fighting over territory. © Christian Schwarz

Indonesian mangrove crab blowing bubbles

Two sparring male California fiddler crabs, *Uca crenulata* (Ingrid Taylar)

Pom pom crab with eggs
© Chatchadaporn Kittisaratham

Tree-hole crab (*Sundathelphusa* sp.) female with brood in a rainforest in Panay, Philippines. This is a freshwater crab like the panther crab, and these related species have large eggs and no planktonic stage. © Christian Schwarz

This male fiddler crab had his enlarged cheliped torn off and carapace dented by a larger male while fighting over a mate.

and most or all of her legs removed. Many crabs and terrestrial hermits hide in burrows for protection when molting. If a male dug up a teneral female for mating, she would be exposed to the world and become an easy meal for predators. Swimming marine crabs have no permanent home, molt in the open, and are known to mate only after molting. An adult male of one of the most familiar swimming crabs, the blue crab, is a vicious, aggressive creature that will chew up a few adult crayfish and fiddlers in a sitting. However, the male is very careful and does not harm the recently molted female. Teneral mating is less likely for burrowing crabs and is probably carried over to few, if any, terrestrial or semi-aquatic species.

Courtship embrace of *Metasesarma aubryi* prior to mating

At the current time, no species of crab has been shown to reproduce through parthenogenesis. The only documented parthenogenetic decapod is the laboratory marbled crayfish (*Procambarus fallax*), the male of which is unknown. Both a male and female crab are needed for successful reproduction. Female crabs may be able to hold viable spermatophores six months or more based on estimates of when females may have molted and produced eggs (Vinuesa 2007). I have had mangrove crab females produce eggs after being without a male for six months and the eggs did not develop. Previous eggs by the same females developed and hatched when a male was

Geosesarma blowing bubbles during mating

The Christmas Island red crab engages in mass migration to the sea. (John Tann)

present in the enclosure. Fiddler crabs do not seem to produce eggs when there has been no recent mating, however they might eat infertile eggs before they are noticed.

The egg membranes are nearly always clear, so development can be observed as ova progress towards hatching. The eggs start out yellow, orange, red, or black, and change to greenish-gray a few days before hatching. The color changes because the shell is transparent and the yolk color is different from the developing larva. The eyes are often visible in the last few days or weeks. Dead and infertile eggs also change color, but usually look different with amorphous coloration and no eyes. Developing eggs swell in size, sometimes doubling. The female crab takes care of the eggs, fanning them with the pleopods, removing foreign matter with the chelipeds, and moving eggs to areas of adequate moisture and oxygen. If the mother dies or the eggs are discarded, there is very little chance for successful hatching. Healthy eggs hatch after two to four weeks (the large eggs of freshwater crabs and some sesarmids can take two to four months to develop). The vast majority of crabs produce thousands, if not tens or hundreds of thousands of eggs, in a single brood. The uncountable numbers of tiny eggs hatch into clouds of planktonic larvae. These require clean, oxygenated saltwater and varying types of phytoplankton and zooplankton to consume as they develop into what will eventually resemble the adult in miniature. Females know when the eggs are about to hatch and return to the sea to release them. Late stage larvae sense their surroundings from within the membrane. If a gravid female with well-developed eggs is moved to better quality or higher salinity water, the eggs often hatch within hours in response.

The most famous crab migration in the world is probably that of the Christmas Island crab (*Gecarcoidea natalis*), but all the large terrestrial crabs (including *Birgus*, *Cardisoma*, *Coenobita*, and *Gecarcinus*) must return to the ocean so their larvae can develop. The timing is not necessarily annual. Various land crabs, including at least four species of *Coenobita* hermits, go down to the ocean to release their young during the new moon. The mechanism for this is unknown, but it is believed this behavior gives the young the best chance for survival since the high water allows them to go out to sea and the low light reduces initial predation (Amesbury 1980). Under captive conditions I have observed dozens of brachyuran and anomuran larval releases which did not seem to be timed to anything specific.

The larvae can go through three to fifteen molts and change significantly through these stages. Many of the brachyuran larvae are released as zoea around 0.3 mm, including the appendages. These are often smaller than hatchling brine shrimp and are too small to make out features with the naked eye (they look like tiny, moving specks). They have a globular shape because the tail is folded underneath the body, but it can be stretched out or used to hold large prey against the mouth. They swim upright and move in jumps or in swerving lines. Anomuran larvae are much larger and look like tiny, see-through shrimp, rarely more than 1.5 mm long. As they swim they appear to float up and down in the water. They swim upside-down with the tail at the top. Unfortunately, even when provided cultured green water (phytoplankton), rotifers, or hatchling brine shrimp according to size, the outcome for captive planktonic larvae is most commonly death. With the correct salinity (usually 1.020) and high water quality, but without adequate food, they die in three to five days as they run out of stored energy. In some species the initial molt or molts do not require any food. Even if available food is adequate, a plankton tank may be needed to keep the larvae from settling on the bottom where they die. Aquaculture facilities using wild collected and strained plankton can successfully rear out tiny crabs in massive numbers, but this is beyond the capability of most aquarists with moderate means. On the other hand, some species produce slightly larger larvae that can be fed with baby brine and these, however difficult, are within the realm of possibility for the devoted, experienced enthusiast.

I remember first trying to rear planktonic crustacean larvae in the mid-1980s using bare-bottom aquariums, airstones for movement, and cultured phytoplankton for food, but within a few days to a week the larvae would always die. More recently I constructed a small plankton tank, because these have been used to rear crustaceans like harlequin shrimp that were not previously successful without aquaculture facilities (marine aquaculture facilities are usually right on the ocean to allow access to fresh ocean water and wild-harvested plankton). My first attempt with *Uca minax* was entirely unsuccessful because the larvae were unbelievably small. I was able to grow *U. pugilator* through four instars because it hatched out significantly larger (yet was still barely visible to human eyes). Hopefully the plankton tank will work with eggs from a species that is not so tiny or that is better suited

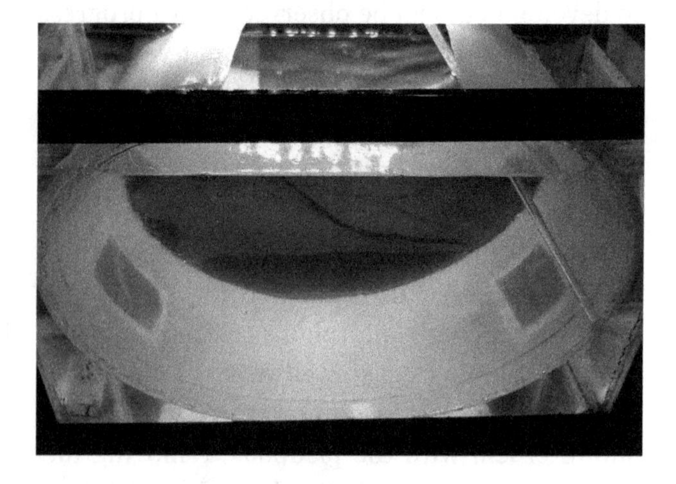

A plankton tank offers a circular flow to keep planktonic larvae from settling in the corners.

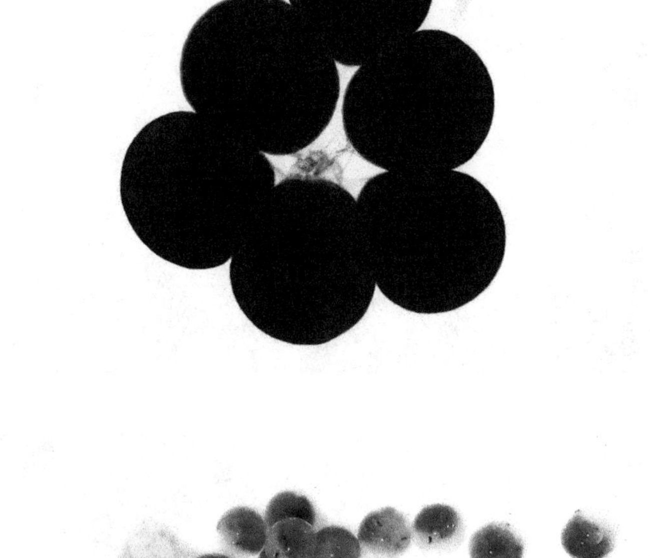

These *TINY* eggs (from *Metasesarma aubryi*) have stalks that connect and help hold together the large mass of eggs held by the female's pleopods.

Healthy *Geosesarma* eggs may be abandoned by the female when she is disturbed. Relative size and development (eyes) can be seen. They were kept alive till hatching, but none hatched correctly.

Large eggs are fewer in number, however this female *Geosesarma hagen* appears to have abandoned most of her eggs.

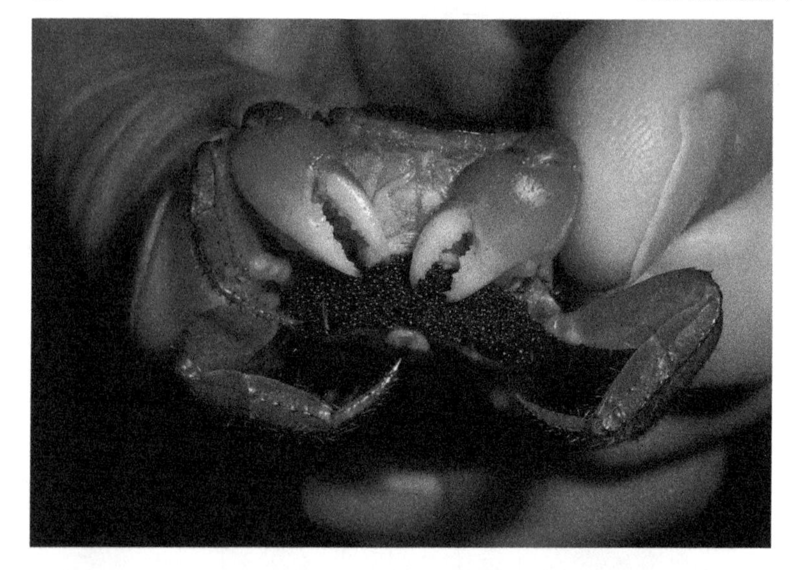

Metasesarma aubryi with freshly laid eggs five days following observed mating. The female molted two months prior and lost a leg in the molt. (The small leg bud seen here recently developed and will enlarge as the next molt approaches.)

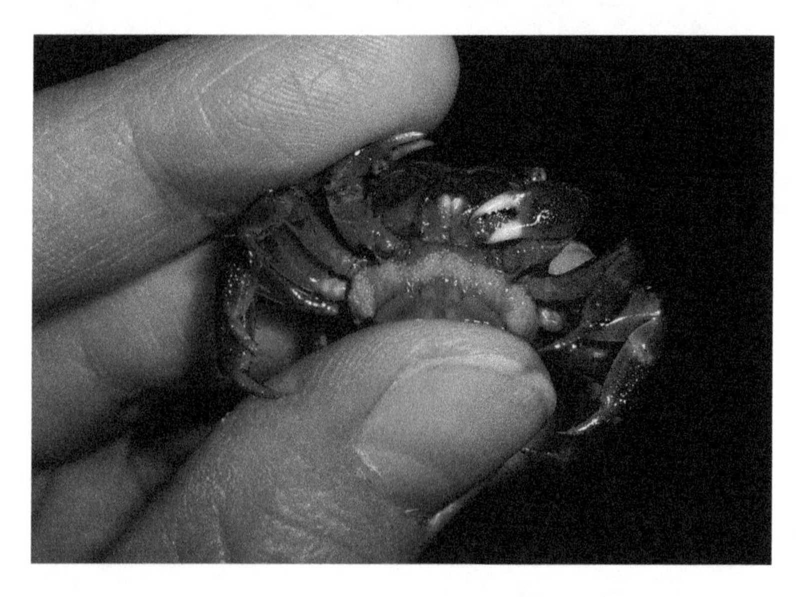

This mangrove crab's infertile eggs turned a strange orange color after a week, May 2016.

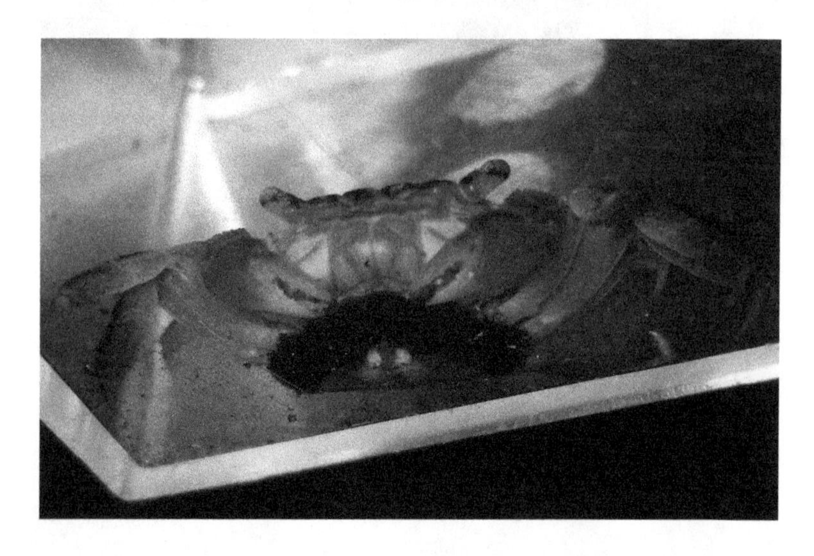

Mangrove crab with well-developed fertile eggs, October 2013.

Mangrove crab egg color of freshly laid infertile eggs, May 2016.

to the phytoplankton and rotifers that are commonly available.

Although nearly all the marine decapods and terrestrial crabs reproduce by way of planktonic larvae, crayfish and freshwater shrimp have large eggs that develop into the "adult" body form. This has allowed a number of crayfish and freshwater shrimp to become popular with aquarists in the last few decades. Unfortunately, all the commonly available brackish water crabs and land crabs must return to the ocean to reproduce, making captive reproduction impractical. Big, colorful "soap dish" crabs and hardy hermit crabs offered handsome, but somewhat dead-end, pets since at least the 1960s. Around 2006, new crabs dubbed 'vampires' entered the terrarium hobby. These species could be reared in captivity by the average hobbyist. Females of these crabs produce large eggs that develop into tiny crabs within the nursery formed by the folded tail and pleopods. Due to the smaller abdomen and need to protect eggs, the number per brood is commonly measured in dozens, rather than thousands or hundreds of thousands. *Geosesarma* eggs are hundreds of times the mass of most planktonic crabs. Each egg is approximately 1.4 mm versus 0.2 mm for planktonic ova.

Direct development is standard for the true freshwater or river crabs, of which there are more than twelve hundred, but only a few species have become available and only in the last few years. Even in Europe where the wild-caught invertebrate variety is significantly greater than in North America, freshwater crabs were seldom seen in pet shops (Werner 2003). The recent popularization of two freshwater crabs, the panther and Matano crabs, may realize a new trend, especially if they prove to be something the average hobbyist can breed.

Growth

Most crabs develop through two unique stages before taking on the adult crab form. The first stage is the zoea, a swimming filter-feeding stage that primarily consumes phytoplankton. It uses its legs (pereopods) to swim and has reduced or absent abdominal appendages (pleopods). The second stage is the post-larva, which is also free-swimming, but it uses abdominal appendages for swimming and has claws. This stage is usually predatory on zooplankton. In true crabs it is called the megalopa, while in hermit crabs the same stage is known as the glaucothoe. The

Blue crab females produce hundreds of thousands of tiny eggs.

This female has just released thousands of hatchlings. She was placed in this shallow container of full strength saltwater a day earlier because the eggs appeared to have developed. A piece of sponge had been in the container to keep her from drowning, but was removed for the picture.

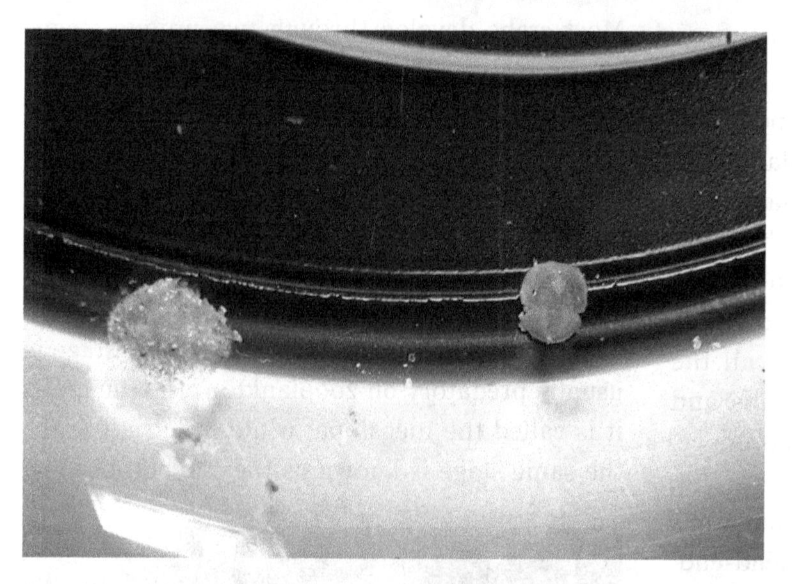

Horseshoe crabs have large larvae (right) that molt to second instar (left) without feeding. These are not crustaceans, but are more closely related to spiders than true crabs.

Geosesarma dennerle immatures a few days old. © Joe Reich

(*t*) Mating captive-reared *Geosesarma tiomanicum* in a tunnel under a piece of wood.

(*b*) Some crabs like this *Geosesarma* female can be difficult to tell when they are holding eggs. In this case a rather large brood causes the abdomen to bulge whereas smaller broods lead to no change in the outward appearance.

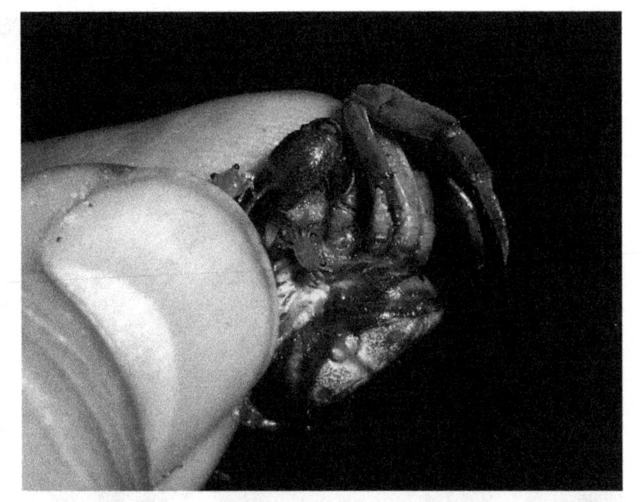

(*t*) Vampire babies are incredibly large compared to planktonic hatchlings.

(*m*) 1st instar *Geosesarma* sp., "orchid"

(*b*) 3rd instar *Geosesarma* sp., "orchid"

This orchid vampire mother was disturbed and began grabbing young from beneath her abdomen and tossing them away. This is likely a defensive mechanism to save babies when the mother is about to be eaten, but the young will die if they are not fully formed.

Freshly released *Geosesarma* are white, but usually darken to gray or tan in a few days.

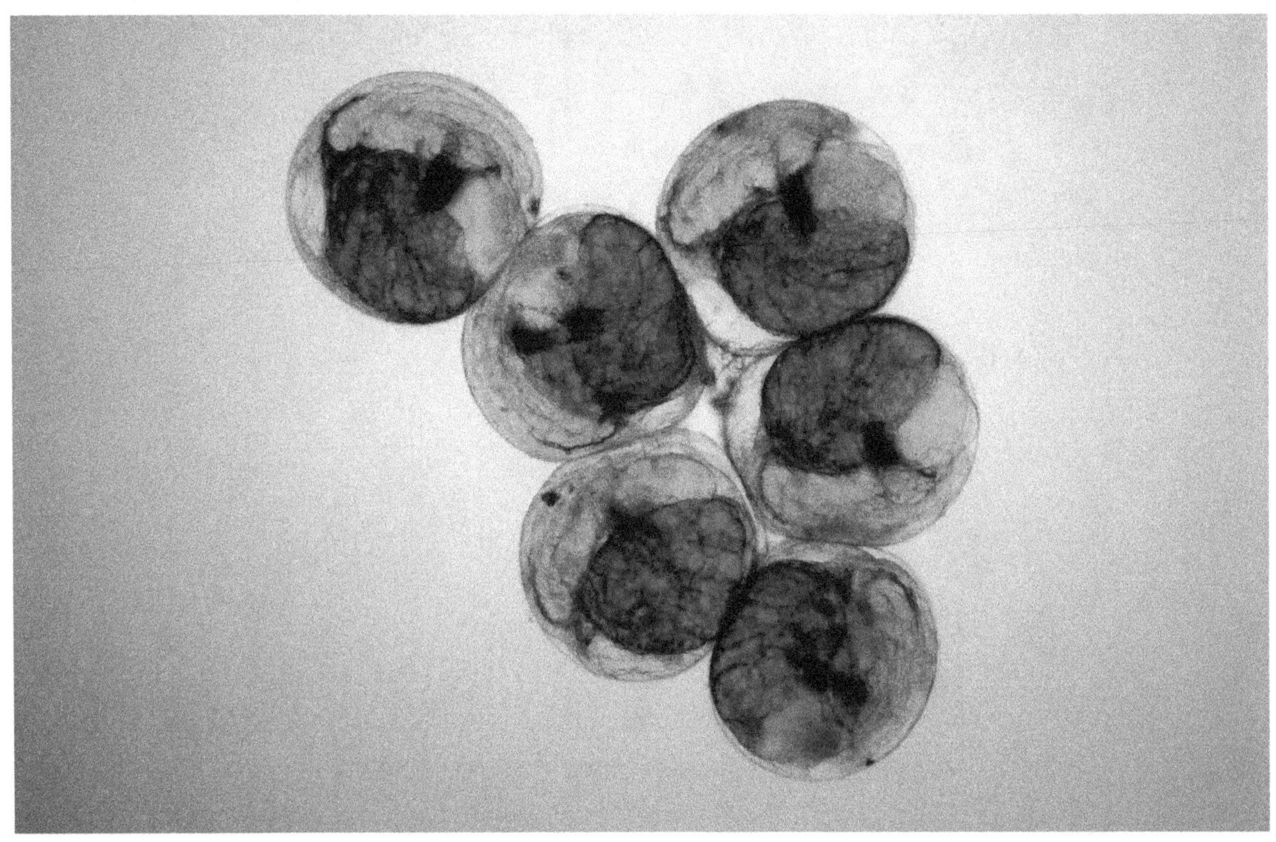

Percnon ova appear red to the eye, but a microscope shows the membrane is clear and only part of the developing larva is red.

These dwarf mud crab eggs are developed and hatched a few hours later.

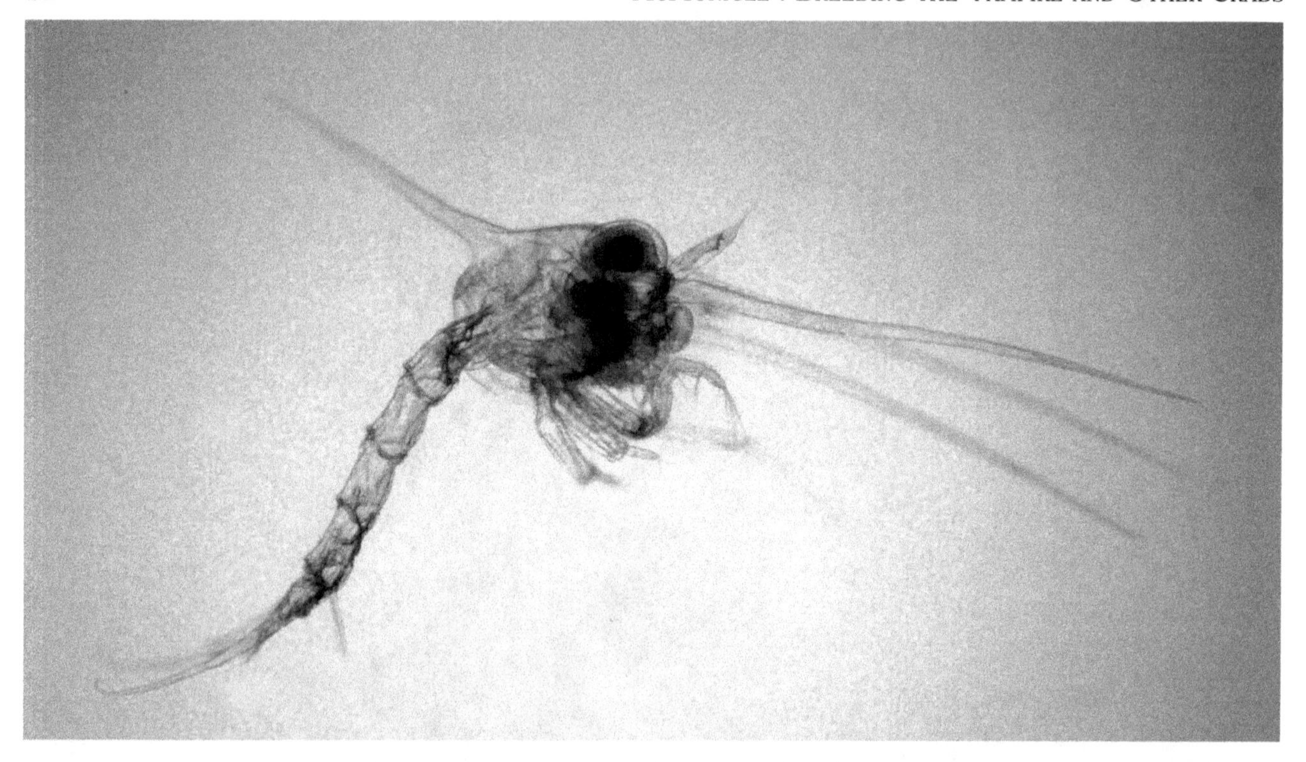

Rithropanopeus harrisii zoea at twelve hours.

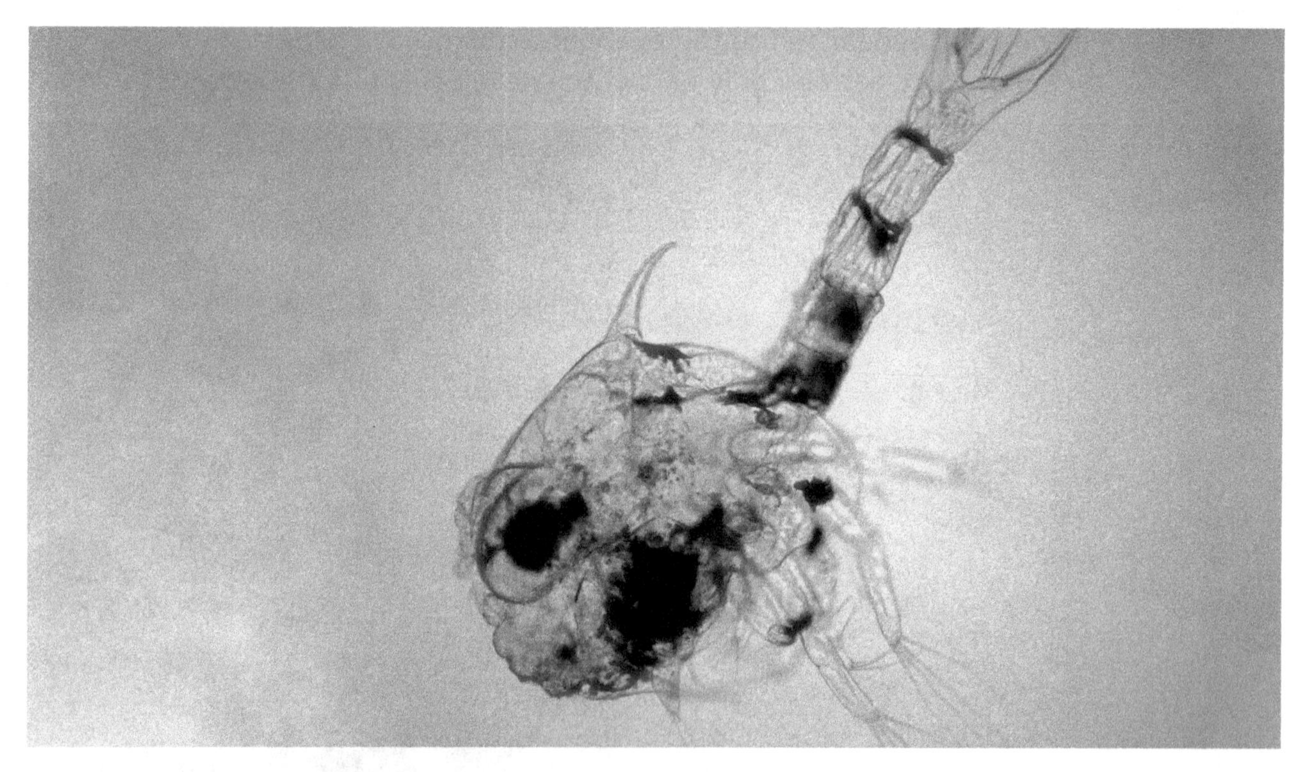

Uca pugilator first instar zoea at twenty hours.

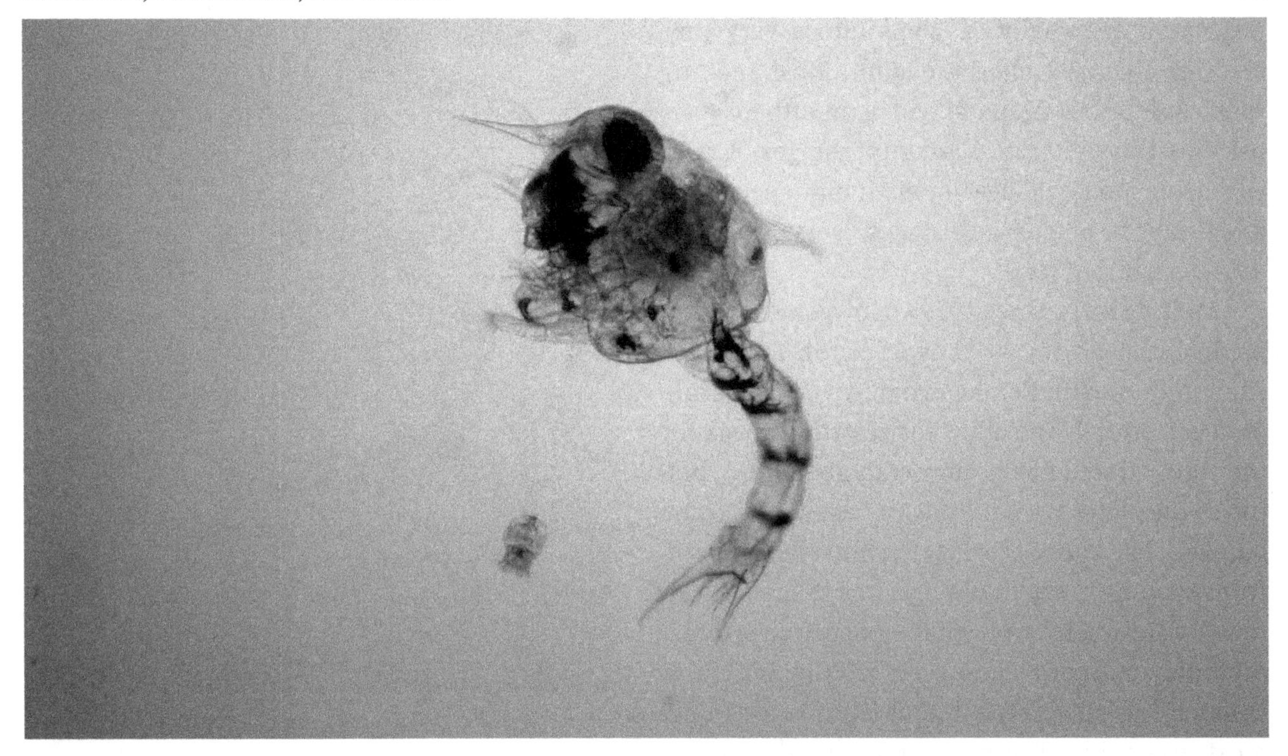

Uca pugilator third instar zoea at eleven days.

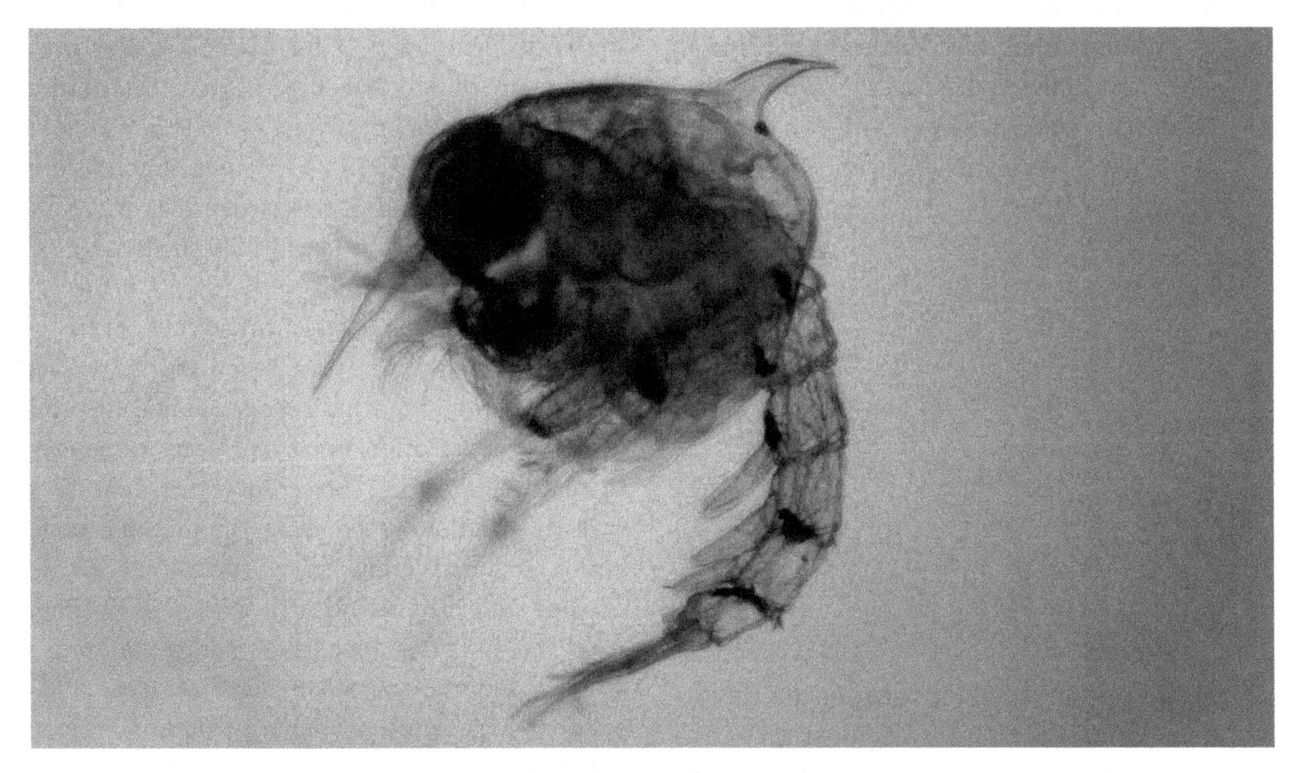

Uca pugilator fourth instar at seventeen days.

post-larva eventually develops into a very tiny creature that resembles the adult. Most anomurans and brachyurans spend a month or two passing through the planktonic stages. Crabs with truncate larval development may spend just a few days in one or both stages, lack a stage, or develop without food.

Molting is the staged growth of all arthropods as they cast off their old exoskeleton for a new shell underneath. Some portions of the exoskeleton can stretch and allow for growth (especially the thin, stretchable hermit crab abdomen), but the main plates do not grow or regenerate from damage. The entire exoskeleton must be replaced from time to time. Young specimens molt every few days or weeks, while mature specimens often molt only once or twice a year. Molting for crabs is much less structured than it is for insects and most arachnids. The number required to meet adulthood does not seem to be fixed. The zoea and post larva stages of the planktonic forms can last a different number of instars according to the species or individual. Adult crustaceans also grow differently from insects and the majority of arachnids because they usually continue to molt after reaching reproductive maturity. The adults can molt and grow much larger after first reaching sexual maturity. Older, larger females produce more rather than bigger eggs. Even more interesting to the enthusiast is that they are capable of shrinking to fit a new environment. Very large wild-caught adult crustaceans which do not get adequate nutrition in captivity shrink significantly at their next molt.

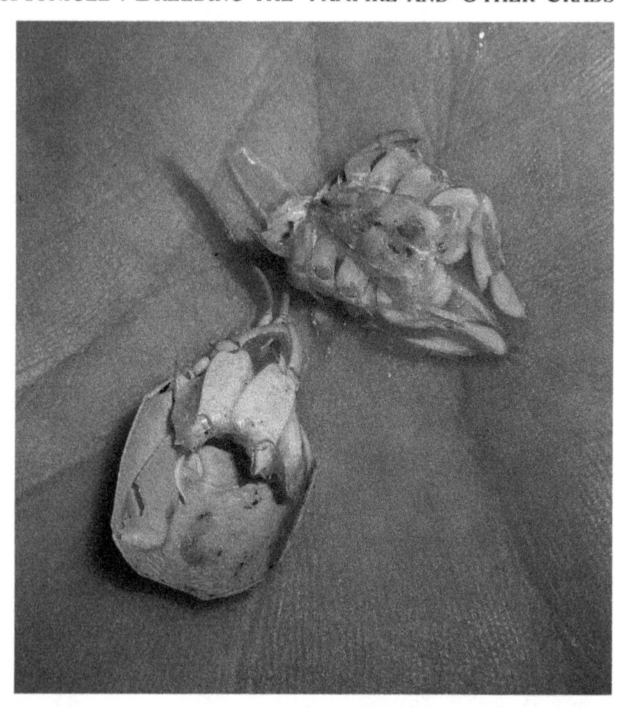

Mole crab exuvium

Geosesarma sp. "orchid" stages of development and coloration. Specimens are from the same brood and are the same age. © Joseph Reich

Some crabs molt every few weeks, grow to maturity quickly, and can reach breeding size in less than ten months. Terrestrial hermits and deep-sea crabs often need five years or longer to mature. Molting starts out every few days to weeks and slows down as the animals grow larger. In many crabs the normal adult size is reached and minimal growth is seen over time. However, for some species the young males and females average much smaller sizes than older specimens. Adult size and maximum size are not necessarily similar for crustaceans. Development and size increases seem tied more to food than temperature, but temperatures that are too

Geosesarma part way through molting.

Geosesarma shortly after completion of molt.

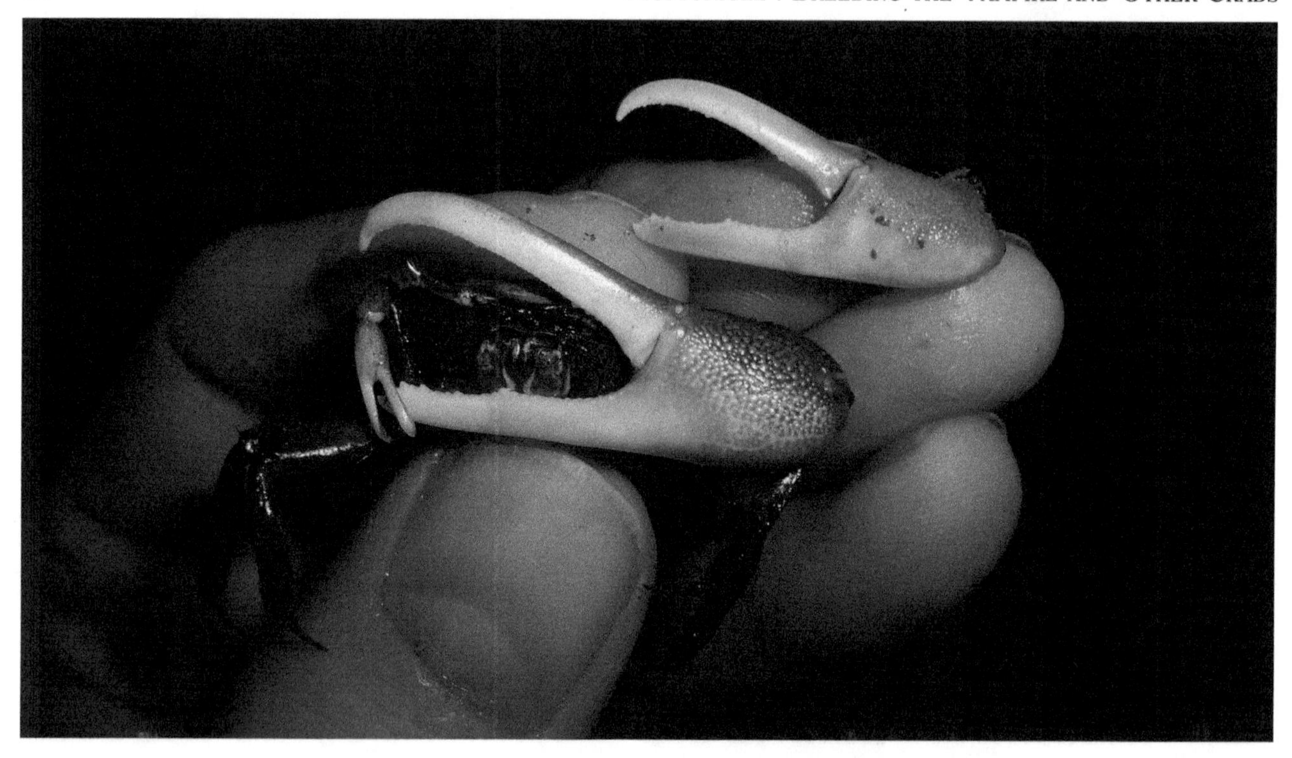

Uca minax, chela size increase in a single molt

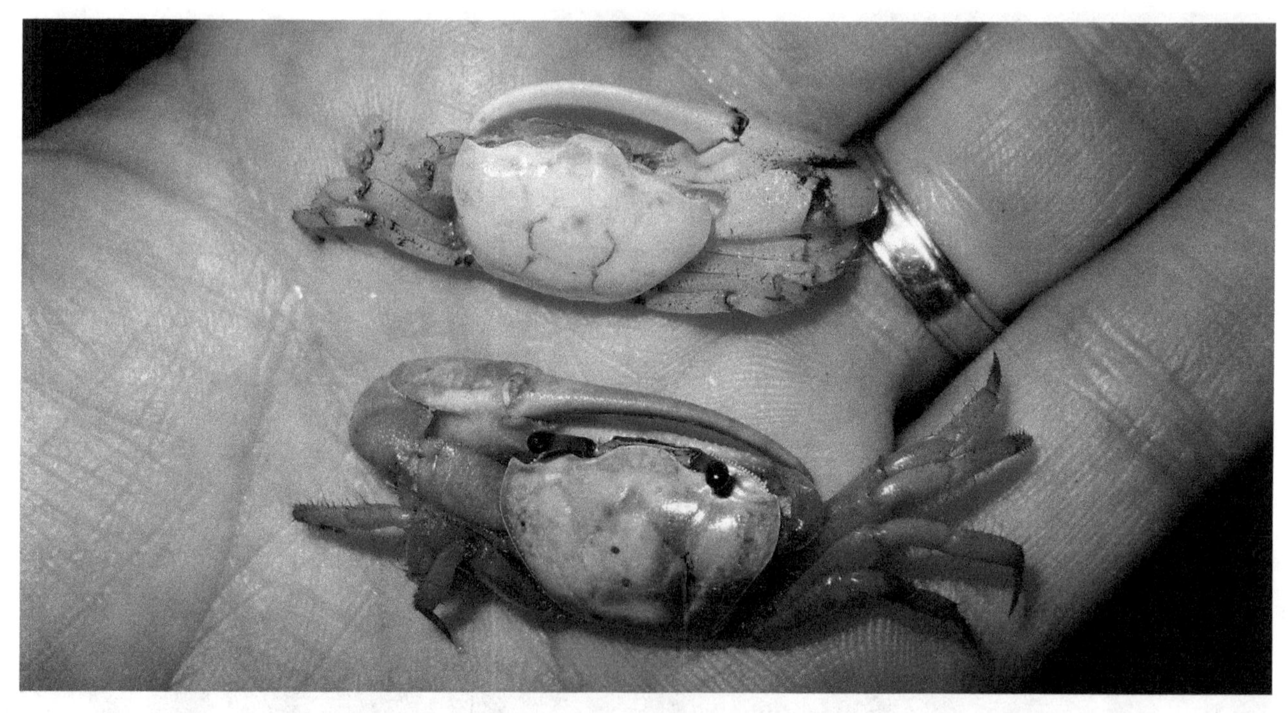

The exuvium (*top*) can look a lot different from a dead crab (*bottom*).

cold can prevent feeding, cause bad molts, or kill them. For tropical species, 'too cold' is less than 72° F (22° C). Without enough food they may take a long time between molts or not molt at all—warmer temperatures will only make this worse since more energy is consumed. Sometimes the adequacy of the captive diet cannot be determined until the animals molt. If they do not regenerate, have rippled exoskeleton surfaces or twisted appendages, or the whole animal shrinks or never grows, the diet is likely deficient. Severe diet issues can result in death during or shortly after the molt.

The discarded molt, or exuvium, is normally eaten by the crab, though not immediately as with many insects and terrestrial isopods. This recycling of materials is far more important for freshwater species and can be crucial for the quality of the next exoskeleton. Many marine species do not return to eat the exuvium, if disturbed, and are not adversely affected if the shell is removed. The exuvium of freshwater species is often very thin and lacks pigment of the living crab (*Geosesarma*) or is so thin it appears translucent (*Parathelphusa*). Marine crabs can be more difficult to tell from their old skins. *Panopeus*

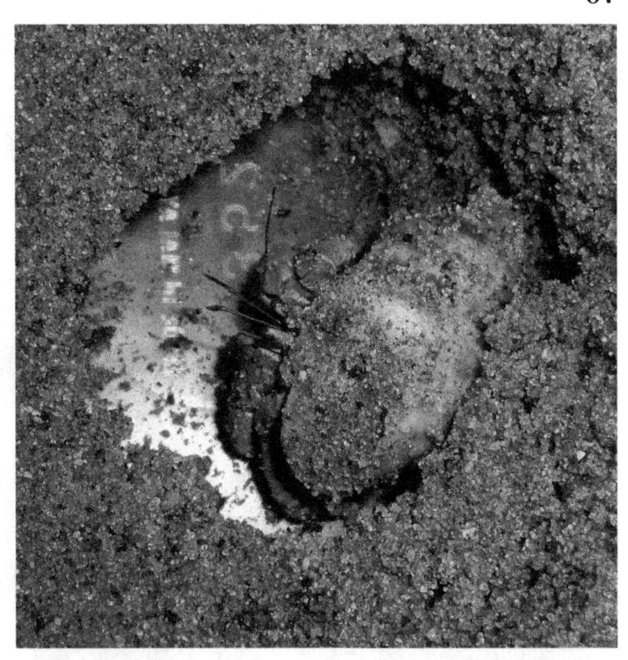

This terrestrial hermit built its molting cell under a stone water dish (*Coenobita clypeatus*).

Molts can look a lot like a dead crab. One of these is a molt (*Somanniathelphusa* sp.).

Molting cell contents following the molt: the teneral crab is pink in color and soft.

Immature *Geosesarma tiomanicum* and cast off shell

Uca minax exuvium and teneral specimen

exoskeletons are very thick and hard and can be difficult to distinguish from the living creature. *Percnon* shells are also thick, but they are easy to discern from living creatures because the tips of the eyestalks are caved in and the tail usually does not stay shut. Fiddler crab (*Uca*) exoskeletons are moderately thin, but they can be confused with a living or dead crab. Sometimes they smell like a dead crab, which is rather unusual for an exuvium. The older (bigger) the crab, the sturdier and thicker the shed exoskeleton becomes.

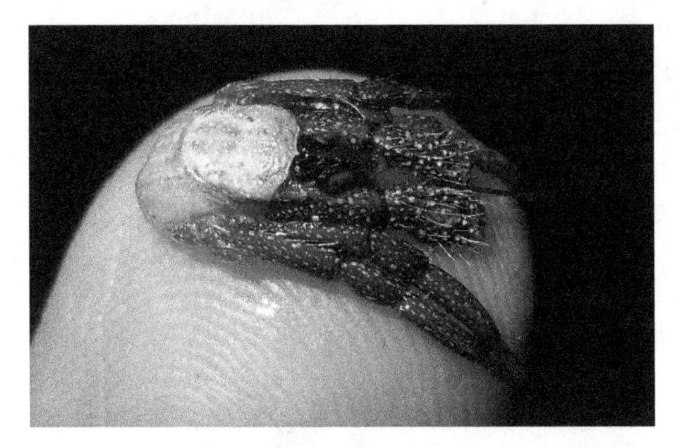

The abdominal exoskeleton of hermit crabs is extremely thin and difficult to locate on the exuvium. Mexican red leg hermits (*Clibanarius diguetti*) molt every few months and do not eat the old shell.

Cardiosoma armatum molt and teneral crab

Regeneration

Crabs shed legs from stress, illness, self-defense, and territorial fights, but they regenerate lost appendages. The walking legs are often close to full-size after a single molt. Sometimes the regenerated leg is slightly twisted or bent at the tip, which will correct itself at the next molt (unless it gets snagged on a decoration and is autotomized). If only the last segment or two of the leg is lost it almost never develops correctly. Walking legs do not always regenerate and a missing rear leg may take a few molts or never be restored. Sometimes the legs do not grow back at the next molt because resources are routed to a missing front leg (the claw or cheliped) and the walking legs will be regenerated with the subsequent molt. The enlarged chelipeds almost always return in one molt. However, a lost cheliped may never return to full size no matter how many molts occur. Over years of keeping true crabs the keeper is likely to see hundreds of legs lost and regrown. Hermit crabs do not autotomize as readily and have less impressive regenerative abilities. Young crabs regenerate quickly and often seamlessly, while older crabs often take longer to molt when damaged. An old crab may never look quite right or may die before there are enough molts to fully correct the problem. Eyes and mouthparts should also be able to recover from damage when the animal molts, but these are uncommon losses. Appendages lost too close to the base, specifically eyes, may never grow back. I acquired a female *Uca* sp. with a missing eye and after five molts the eye has still not grown back.

Crabs do not require a molt for limited regeneration. They normally grow back a leg bud (a folded, tiny, soft version of the leg) without undergoing a single molt. Unless the next molt is imminent, crabs grow back the bud inbetween molts and then regenerate a fully formed leg at the first molt. The bud usually swells and grows without a molt and can be as much as a

Lost cheliped with no sign of regrowth (*Gecarcinus*)

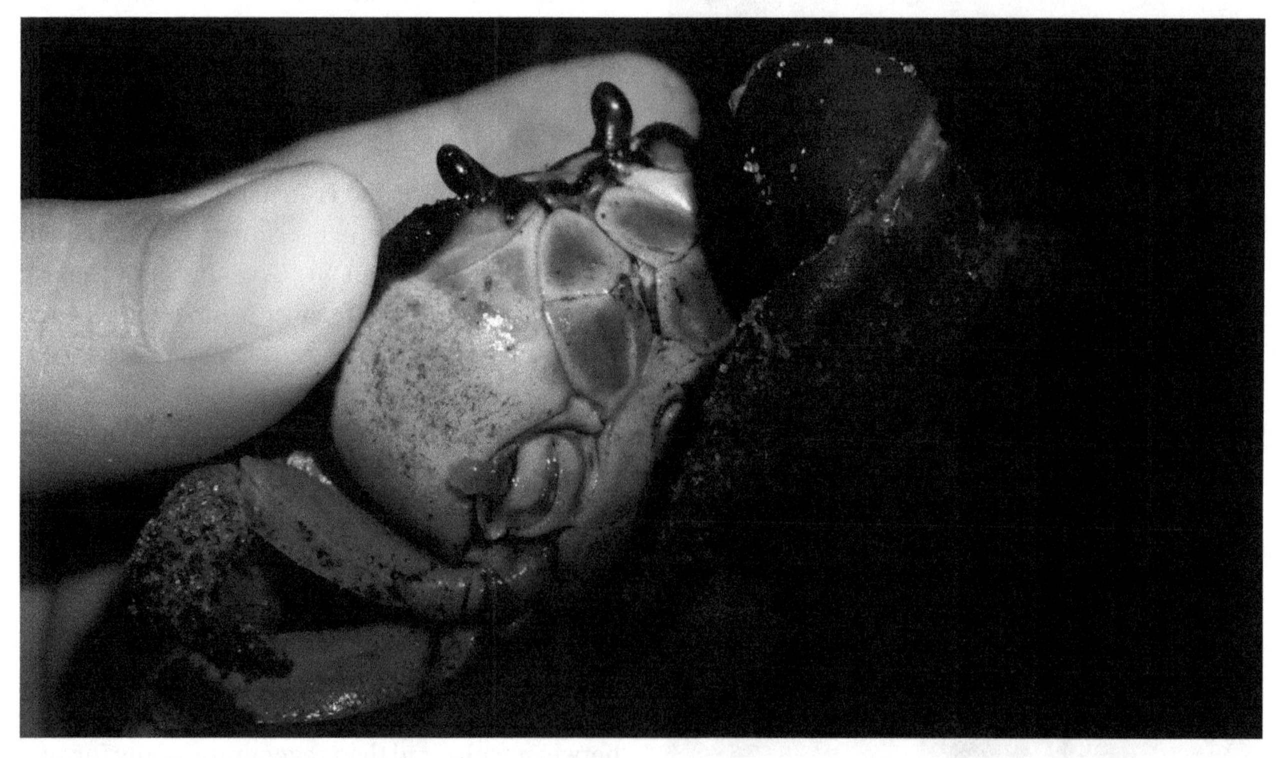

The cheliped beginning to develop without a molt (two months later)

third the mass of the leg just before the exoskeleton is shed. This proceeds differently from regeneration in other familiar arthropods. Insects and arachnids often do grow back a tiny leg bud before the full version is regrown, but only following a molt, never before. This means insects and arachnids often need two or more molts to regrow a functional leg, whereas crabs need only one molt.

While many types of crabs have both "right-handed" and "left-handed" forms, the difference can be fixed for some groups. It may even be used to determine orders, specifically for marine hermits. In some crabs the large claw can switch sides on the same individual. Male crabs with a large and small front claw, specifically fiddlers and patriot crabs, sometimes lose the enlarged claw to a predator or rival. The smaller claw is used mostly for feeding so the loss of the big claw does not usually hinder survival. When the big claw is regenerated, a tiny feeding claw becomes its replacement. Though smaller than the original, the previously smaller cheliped on the other side grows back as the enlarged claw.

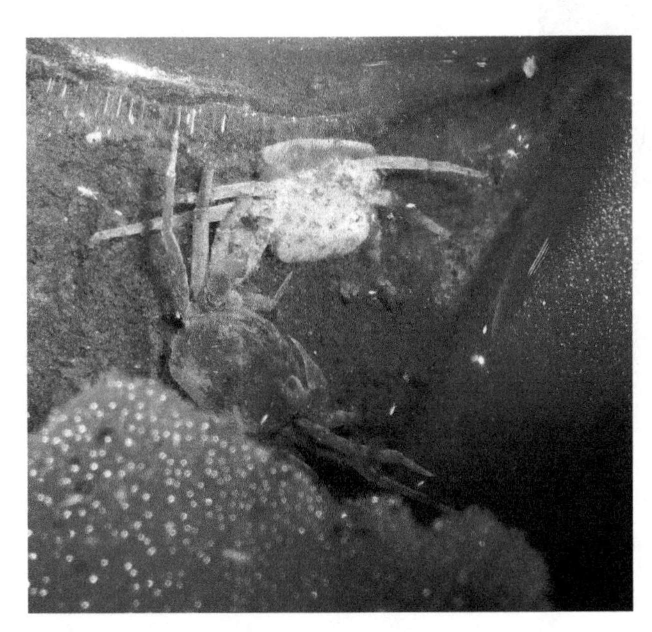

Geosesarma dennerle: discarded exuvia may stay untouched for a few weeks.

Longevity

Captive experiences and estimates based on surveys and tagging of wild animals suggest individual crabs can live a few years to a few decades,

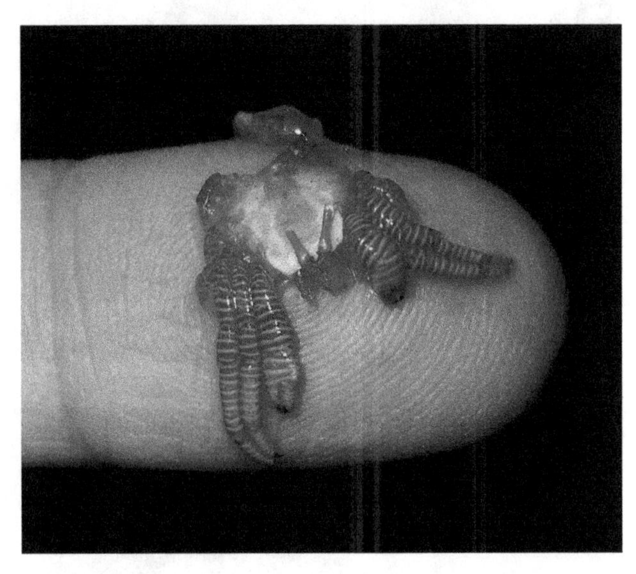

A small specimen of the marine *Ciliopagurus* hermit casts off a very thin and flexible exuvium.

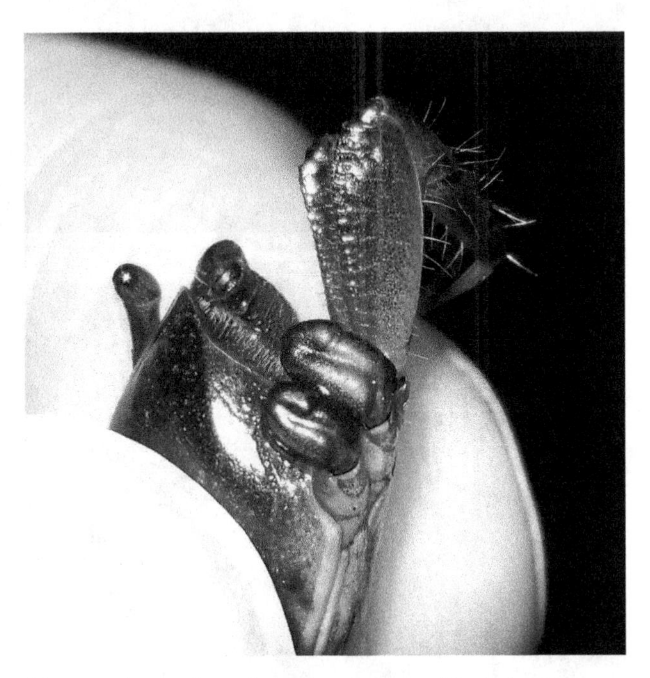

Uca minax leg bud regeneration before a molt has taken place.

The exuvium is often very thin and pale (orchid vampire).

The cheliped and walking leg have grown back as small buds without a molt, *Cardisoma armatum*.

This missing claw from a fight will grow back, but the animal should be separated.

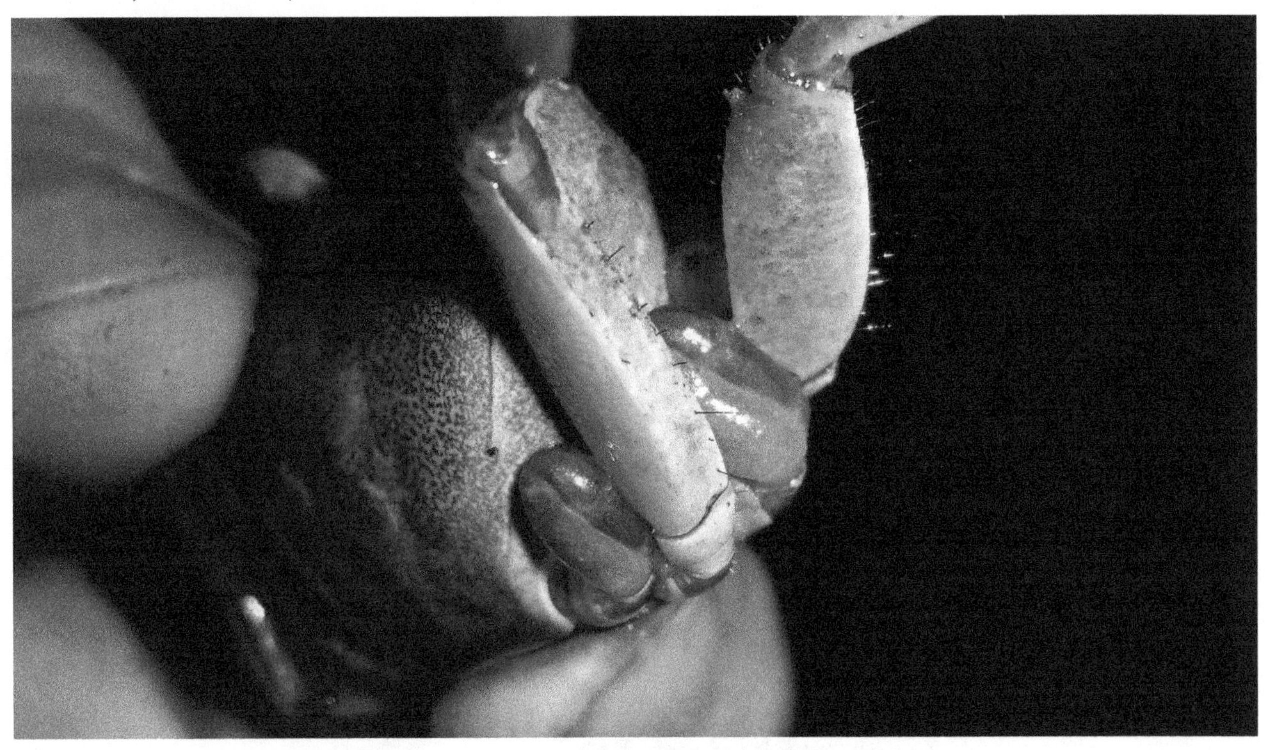

Leg buds have grown back without a molt, *Cardisoma armatum*.

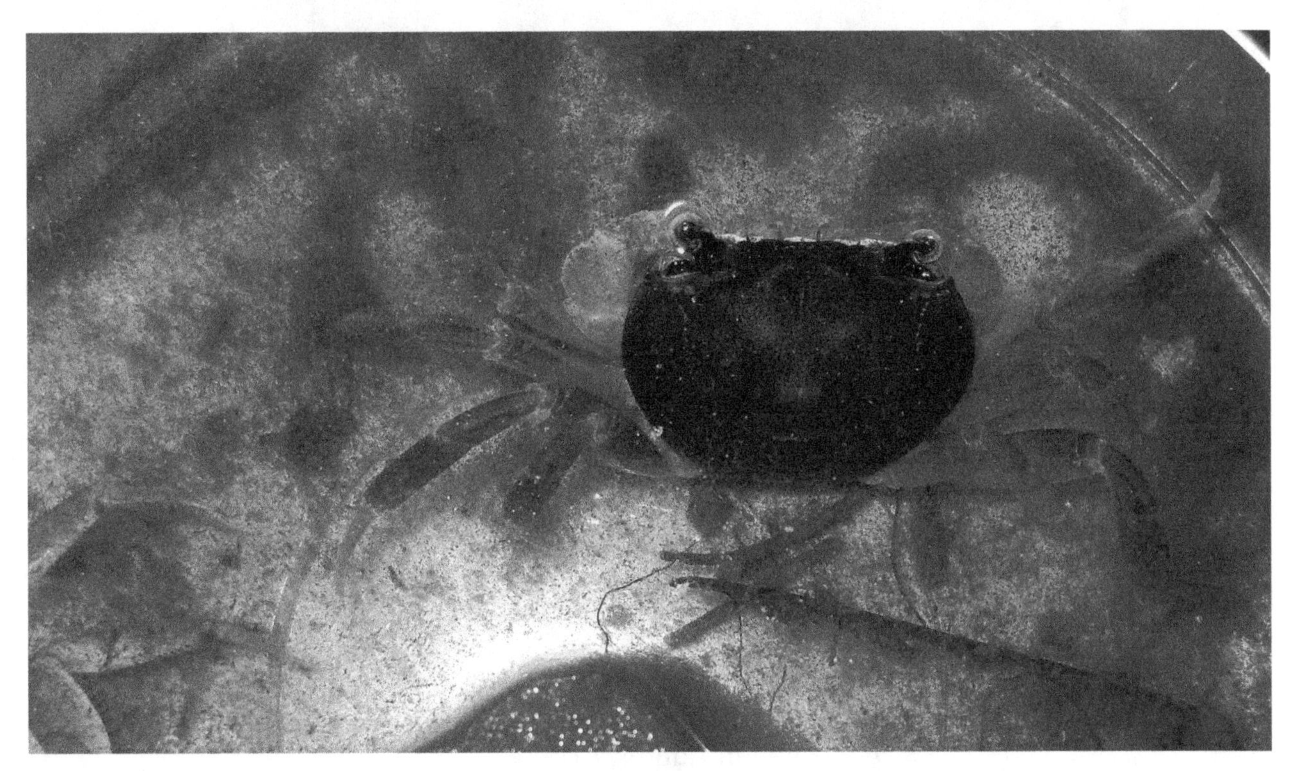

After a single molt the leg buds have grown into fully functional appendages. The regenerated cheliped and two walking legs on the right side are extremely pale following the molt. Regenerated legs are often a different color until the subsequent molt.

depending mostly on the species. Anomuran crabs currently hold the longest records. One hermit enthusiast, Carol Ormes, claims the longest-lived hermit crabs surviving to 37 and 41 years of age (both wild-caught *Coenobita clypeatus*). King crabs can live about 25 years, including the eight years taken to grow to full size (Ingle 1982). Deep-water brachyurans like snow crabs live more than a decade, but about seven years is spent reaching maturity. Land crabs, *Cardisoma* and *Gecarcinus*, have been kept in captivity about as long as hermits, so there is some documentation of their longevity under captive conditions. Both commonly survive up to ten years after they are acquired as full-grown adults. *Geosesarma* species reared in captivity have lived only four years, including the time taken to develop from hatchling to full-grown animal. Fiddler crabs are similar, rarely living more than three years as adults. One commonly kept group of marine anomurans, the porcelain crabs, seldom live a year in captivity unless they are acquired as small crabs.

This blue land crab (*Cardisoma guanhumi*) is missing legs at Pelican Island National Wildlife Refuge. This animal may already be a decade old. (Keenan Adams / USFWS)

HUSBANDRY

Artificial Habitats

Captive enclosures are almost as varied as the crabs themselves. However, the two main types are fully aquatic and semi-aquatic. Fully aquatic requires active filtration, whether a sponge filter, undergravel, outside power filters, or trickle filters. Nearly all crabs are hardy compared to other aquatic creatures, but big crabs usually have even bigger appetites that can lead to rapid fouling. Filters can clog weekly and water changes may be required a few times a month. The problem with aquatic crabs is that most species are escape artists—the wires and hoses needed to keep them alive can lead to their escape and rapid demise. There are limited caging options for full aquatics because filtration is necessary to keep the water clean or the crabs die. Seashore caging that approximates high and low tide can be used with shore crabs, but a daily cycle of high and low water levels is unnecessary for survival. I have kept mangrove and fiddler crabs for years in cages with sponge islands just like those I use with vampire crabs. Many species are destructive and energetic, so it is difficult to decorate the enclosure with live plants.

There is debate whether crabs do better in a decked out "ecosystem" terrarium or a utilitarian rearing container. Perceived "happiness" of the crab is often placed above survival and health, but happiness is often judged by human standards. While the effects of certain stresses may be observed, feelings like "lonely" and "bored" are often attributed to crabs because the keeper thinks they might feel a certain way if they were the crab. In most cases, it is harmful to keep "social" crabs together, even if they only kill or maim one another on rare occasion. Heavily planted and decorated cages usually make it difficult to observe and respond to health issues or poor feeding. Anthropomorphizing pets is normal and often harmless, but it can lead to harmful choices.

Water Quality

Water quality refers to many aspects of water chemistry. To begin, where should the water come from? The source determines the dissolved solids it contains. Unsoftened tap water is normally used because it is clean, easily accessible and has a minimum amount of hardness (usually at least 120 mg/L). Tap water may contain free chlorine, and should be allowed to age for a few days before use. Free chlorine found in many municipal tap waters dissipates with time—this can be sped up with an airstone or even more quickly with dechlorinator (available from any pet store). Bottled water from the local store may contain tap, distilled, deionized, or spring water. Spring water is usually better for crabs than tap, because of higher dissolved hardness,

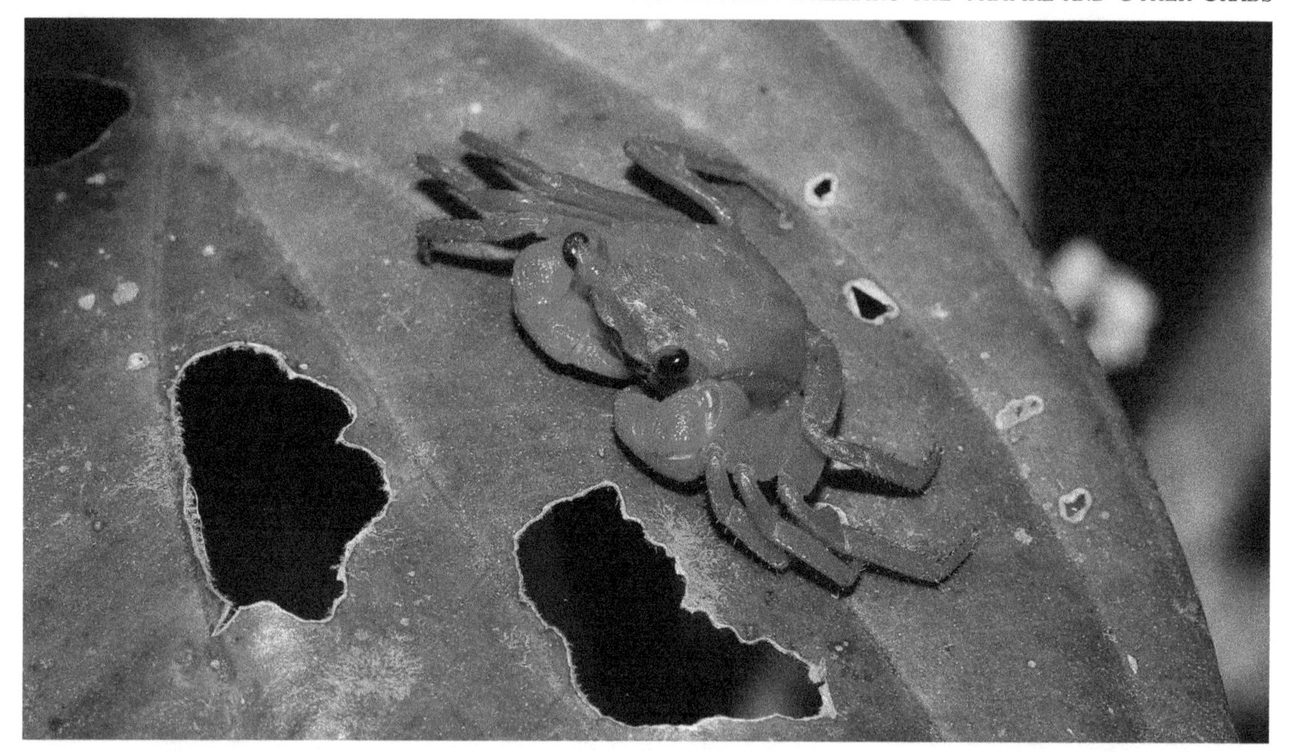

Geosesarma sp. on a leaf in the rainforest, Panay © Christian Schwarz

In its natural habitat in the rainforest, this vampire crab lives in crevices in the bark of large trees, Panay, Philippines. © Christian Schwarz

Orchid vampire adult. Housing does not require dirt or plants. © Joe Reich

Mangrove crab and frass, many species produce a lot of frass. Unlike mammal and reptile feces, crab waste usually does not smell and has similar properties to soil.

This *Geosesarma* sp. is found among limestone rocks on the forest floor in its native habitat in Panay, Philippines. Though some *Geosesarma* are fully terrestrial, others live between water and land on the banks of streams and rivers. © Christian Schwarz

Four panther crabs were reared together, from tiny crabs to mature adults in this 120-gallon aquarium.

Shallow water features are important for most commonly kept species. This *G. tiomanicum* is a recent import that dropped a few legs.

while distilled and deionized are both potentially dangerous because they have had all the hardness and alkalinity stripped. Saltwater mix should be a commercially available grade sold for marine aquariums, never table salt. When mixing a batch of salt water, deionized or distilled water is often recommended because impurities are avoided and the salt mix contains the appropriate hardness and alkalinity.

Crabs may require fresh water, brackish water, of full strength sea water (+ or - 1.020 specific gravity). In the case of the more fully terrestrial species a bowl containing fresh water and another with sea water is suggested. In aquariums and paludariums, it is very difficult to separate salt and fresh. When it comes to salt content, crabs can be very forgiving. Blue crabs (*Callinectes sapidus*) have been reared in freshwater ponds by North Carolina State University. Some marine crabs may accept low salinity if adapted as immatures, but wild-caught adults and large juveniles may not be able to adapt. I attempted to keep blue crabs in extremely low salinity and all three expired within a day or two. They may only be able to adapt to lower salinity when very small. On the other hand, I have had better luck keeping *Uca* sp. in fresh water than in brackish.

With semi-aquatic crabs, it is possible to safely house animals without filtration. Filtration versus no filtration relates to the danger of escape versus the danger of death from overfeeding. Even with filters, the danger of decomposing leftovers is not eliminated. Most filters speed up the process of denitrification where ammonia and nitrite are consumed by bacteria and turned into relatively harmless nitrate. Water quality can be improved by filters, but even oversized filters can be overloaded by too much waste. Removal of uneaten food and use of clean-up crews is suggested (clean-up crews are smaller, often tiny, animals that eat scraps).

Water changes are important for fresh and salt water and with or without filtration. In deep water a siphon hose (usually 1/2" or 13 mm diameter) can be used to remove leftovers and 5-10% of the water at least once a month. Airline tubing can be used to siphon detritus from small containers or cages. A turkey baster is a fantastic investment for removing large amounts of shallow water (compared to an eye dropper).

Escape

The most important aspect of caging is eliminating avenues for escape. Even terrestrial species dry up and die after a few days of exposure in a home. Aquatics may be irreparably damaged in an hour. Many aquatic crabs are nearly impossible to contain in a standard aquarium with power heads or outside power filters, especially the swimming crabs. These crabs swim to the surface to reach a nearby edge or climb cords out of reach from the bottom. Some aquatic species that do not have paddle-shaped rear legs can swim surprising well by moving their legs in a wave pattern, and can be just as difficult to contain. Any outside power filters, including protein skimmers and diatomaceous earth filters, require large tubing that provides easy avenues for escape. Sponge filters or undergravel filters with low air lifts are better options. The water level should be kept a few inches or more (> 50 mm) from the top edge of the aquarium or tub. Pump cords and airline should be hung from an overhead hook above the middle of the aquarium. If cords or tubes are draped over the edge, they can be used for escape. Some aquatic species like arrow crabs and most marine hermits prefer not to leave the water and may even avoid easy avenues for escape. Others, like panther crabs and mud crabs, may seem content for up to a year and a half in the same enclosure and then one day go on a walkabout. Sometimes this is because an originally small animal has, after a recent molt, grown large enough to escape.

Terrestrial crabs also climb cords, tubes, cage decoration, and each other to get out of an artificial habitat. Planning a terrestrial enclosure usually does not require consideration for tubes and cords, but deep substrate offers unique difficulties with containment. Digging can lead to very tall piles higher than the sides of the cage, particularly in smaller caging. Even when there is relatively shallow substrate, a terrestrial hermit crab can escape habitats with walls higher than twice the length of the outstretched legs. Patriot crabs can occasionally get out of an aquarium that is nearly double their legspan, including the top of the highest decoration (this happens at night, so the method of escape is uncertain). Secure lids should be used whenever possible.

Airline should be up tied in the middle of the water, rather than laid over the side of the aquarium, to prevent escape.

Crabs use tiny imperfections in plastic molds and each other to escape enclosures.

Crabs like to hide under things, so finding an escaped crab is very difficult. Aquatic stone crabs and mud crabs end up rapidly dried out

Crabs cannot climb smooth glass, but they are surprisingly adept at escaping enclosures.

Play sand is an excellent burrowing material as long as it remains damp.

in heating ducts as they slip under air registers like slipping under rocks. Swimming crabs may find their way inside bags. Hermit crabs hide under furniture. It is important to find escapees as soon as possible. Even when found, escapees may seem fine but die by the next day since the gills are irreparably damaged from desiccation. Placement of wet towels, water, or food dishes may seem like a good idea to bait an escaped crab, but I am not sure it has ever worked. The best method is to walk around in the middle of the night and early morning with a flashlight (a red flashlight might be helpful if the crab can run faster than you).

Substrate

With aquatic species in small setups it is usually safest to avoid substrate. Detritus should be removed every week or two with a siphon (airline or larger hose) or turkey baster, dropper, or transfer pipette. Gravel or sand will hold a bit of waste and will make it difficult to clean smaller water features. Larger aquariums and paludariums can have under-gravel filters that push water through the gravel, but care must be taken to ensure crabs do not find their way under the filter plate, where they will starve or become pinned while molting. In a large aquarium with a sponge or outside power filter, a gravel cleaner siphon can be used to keep the substrate clean, but deep substrate can still result in anaerobic areas if water does not flow through it regularly.

The exact depth resulting in oxygen depletion depends on the size and compaction of the substrate, burrowing creatures if present, water flow and oxygen levels, temperature, and salinity. Fine sand in warm saltwater can become anaerobic in less than half an inch (< 1.3 cm), whereas coarse coral sand with bristle worms may be okay for a foot. Warm temperatures and high salinity reduce the water's ability to hold oxygen. The bacteria in anaerobic environments (very little to no oxygen) break organics down into methane, carbon dioxide, and hydrogen sulfide, which can be harmful to the water chemistry or directly damage the crabs. It is usually not obvious until the substrate is dug through. Anaerobic bacteria coats everything in deep black and the gases produce horrendous smells. Even if undisturbed it can be harmful to the crabs since the gases build up and release on occasion.

Semi-aquatic species can be provided a layer of peat or coconut fiber to build tunnels. A thin layer, 0.5″ to 1″ (1.2-2.5 cm) is suggested. The substrate should be on the land areas. The land structures can be created with pieces of thick sponge or built on a support rack made of plastic egg crate covered with nylon window screen.

Isopods are commonly used as clean-up crews for hermit crabs, but they are susceptible to drying out.

Large shells make great hides for fiddler crabs.

Different species of *Geosesarma* can survive for long periods in the same enclosure with little to no aggression. The two specimens on the left are begrudgingly sharing a cricket.

Water-logged deep substrate eventually becomes anaerobic or septic due to lack of oxygen. *Geosesarma* are one of the very few that can do just fine if the water quality is terrible. Mud crabs and fiddlers can spend a lot of time out of the water safely, but they still die if the water is fouled.

Terrestrial crabs, moon crabs, and terrestrial hermits should be provided substrate to build tunnels. Damp sand can be a few feet deep without anaerobic bacteria growth if it is not water-logged. If too much water is added to substrate, it can become anaerobic within a few inches of the surface. This usually does not bother terrestrial crabs unless they build a molting cell near or within the anaerobic area. Earthworms can be very useful to keep the substrate aerated, but many species found outdoors die and disintegrate when the temperature gets above 75° F (24° C), which is near the low range for keeping most crabs happy.

A hygrometer can be used to measure the humidity level above the substrate for terrestrial habitats. (The humidity levels at the top of the cage or outside the cage are not that useful.) The humidity should not be below 50% because the gills of "dry" terrestrial species (land hermits and moon crabs) can dry out even if there is a water bowl in the cage. This is usually only a concern in homes where winter heating drops the room humidity down below 5 or 10%. Also, if the surface of the substrate is checked for dampness daily, a hygrometer is not necessary.

Temperature

"Room" temperature 72-75° F (22-24° C) is a great starting point for most tropical crabs, but of course some areas in homes may be much cooler or warmer. 70-84° F (21-29° C) is a practical range for nearly every species the hobbyist is likely to come across (except for living cold-water crabs from the food market). Tropical species can survive and molt successfully as low as 67° F (19° C), but early instar young and very old animals are prone to die at this temperature. Lower than 67° F (19° C) can be deadly to some tropical crabs. 84° F (29° C) is often touted as ideal, but it is not necessary and can increase problems with humidity and fouling, both of which can be deadly. 75° F (24° C) has proven ideal for breeding and rearing vampire crabs, but there is no perfect temperature. Slightly cooler or warmer may not matter at all. The optimum can change with varying aspects of the habitat enclosure or with different food. A basking light or radiant heater for terrestrial and semi-terrestrial species can allow them to choose their own temperature.

Species from temperate waters will grow and molt at room temperature, but are capable of surviving far colder ranges as long as the water does not freeze. Cold-water species often die if kept above 65° F (18° C) and require a refrigeration unit to keep the water cool.

Lighting, Plants, and Decorations

Day and night cycles should be provided but direct lighting is not necessary. Adult crabs are almost entirely nocturnal and prefer not to be active when the lights are on. Planktonic larvae should not be provided night cycles since the limited depth of most home aquaria means they sink and stick to the bottom of the cage when the lights go out.

Lighting is important for maintaining plants in the terrarium. Strong lighting causes wastes to be converted to algae rather than harmful ammonia to a limited degree, even when plants cannot survive with the crabs. A variety of LED lights are available for direct placement on small terrariums that promote photosynthesis and produce little heat. Fluorescent hoods are available for standard aquarium sizes. Household light bulbs in clip-mounted fixtures are an economic alternative that is useful with large caging. Halogens

Coconut hide

New enclosure for moon crabs, before burrowing

Aged enclosure for moon crabs, after months of burrowing

Entire books are written on plants for aquariums and terrariums. This Malaysian gametophyte fern survives fresh or backish and can survive with many types of crabs.

Emergent areas and water features should be easy to climb in and out of.

This *Geosesarma dennerle* is hiding in a tunnel made in a shallow layer of peat on top of sponge. (Java moss has grown over the emergent earth.)

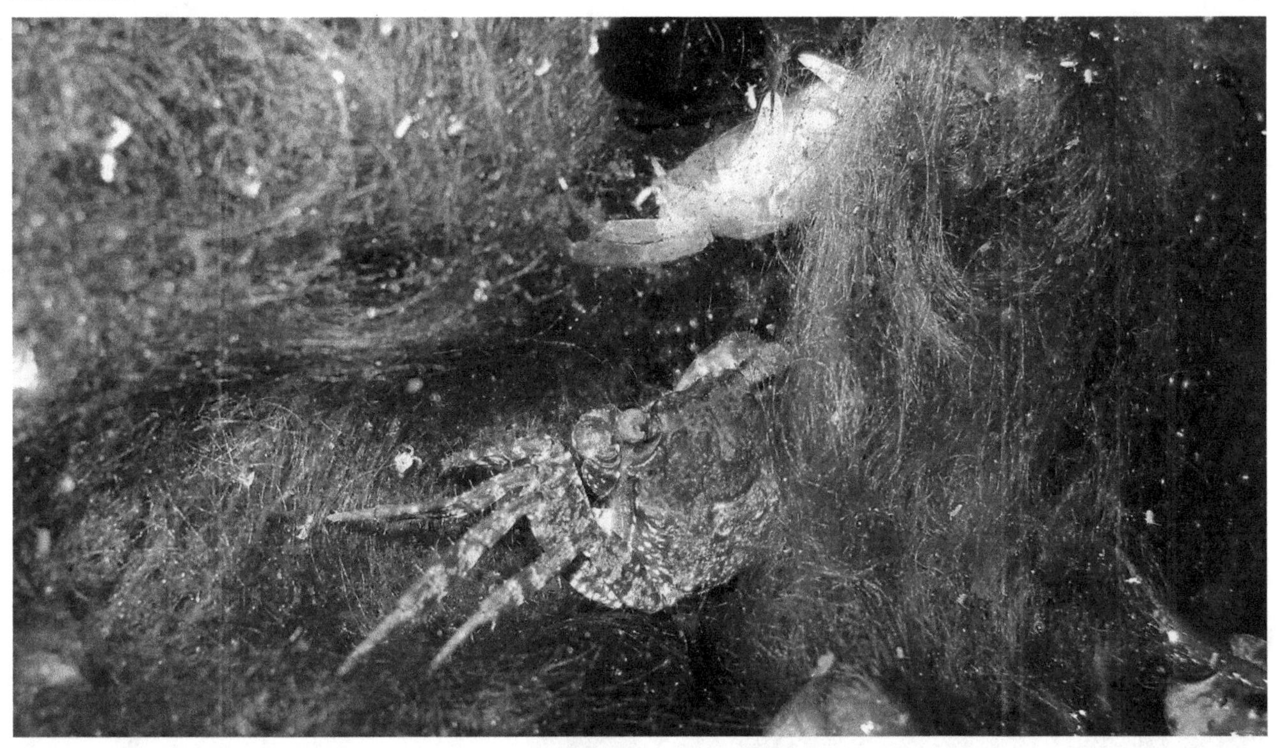

A large immature orchid vampire and its exuvium among hair algae in a shallow water feature. Molts usually take place under water.

may be used as a basking lamp for terrestrials, but the heat produced can cause excessive evaporation or burn plants and melt plastic caging in close proximity.

Though not always practical, vascular plants, macroalgae, moss, and liverworts can be great additions to a crab enclosure. Bromeliads (mostly those from the genera *Aechmea, Cryptanthus, Guzmania*, and *Neoregelia*) are commonly included in the *Geosesarma* and *Metasesarma* terrarium. The crabs will use the bowls of water created by the leaves. In nature these crabs are more often found in limestone or wood crevices—bromeliads are from another continent and are not typical for the habitat. A variety of other vascular plants and ferns can combine to make an attractive enclosure worthy of placement in the living room. Most moss and liverwort survive a narrow range of moisture where they dry up without constant moisture, but cannot survive continuous submergence in water. Most are sensitive to salt and calcium buildup from long term evaporation and replacement of tap water. However, they thrive in terrestrial caging and offer a prehistoric look. Java moss (*Taxiphyllum barbieri*) is a universal terrarium plant because it grows submerged as well as above the water line. It is branched and frilly underwater, but looks like normal terrestrial moss when it grows above water. Java moss is also useful because it can survive some level of salt in the water. Macroalgae, especially *Caulerpa* spp., decorate underwater marine cages and few crabs will eat them.

In addition to or in lieu of substrate for digging, sponge, coconut shells, pieces of wood, cork bark, and rocks can be used to provide shelters for individual crabs. Volcanic and limestone rocks and hole-riddled cork bark can hide young from hungry adults, but it can also prevent the keeper from observing and responding to losses.

Geosesarma hagen, Java moss covered substrate, and a clean-up crew of large white springtails.

Liverwort and mosses growing on a sponge island.

Mosses growing on a thin layer of dirt on a sponge island.

A long-term *Uca* cage is unlikely to support plants (the moss clump is a dead decoration).

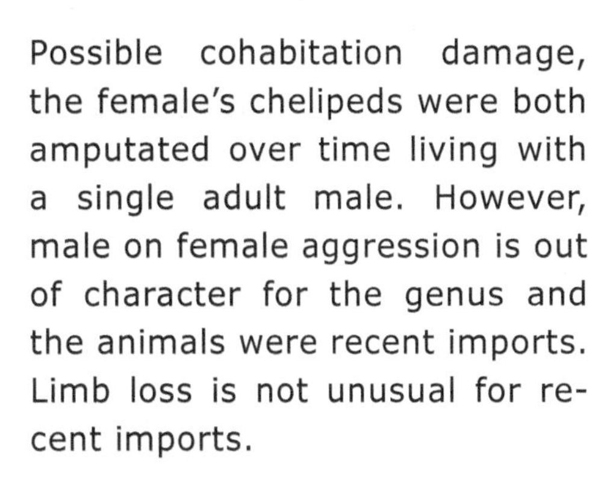

Possible cohabitation damage, the female's chelipeds were both amputated over time living with a single adult male. However, male on female aggression is out of character for the genus and the animals were recent imports. Limb loss is not unusual for recent imports.

Sensitive plant (*Mimosa pudica*)

Water dishes are usually provided for land crabs and hermits. A freshwater dish is a necessity, while a saltwater dish may aid long-term survival and can provide a release site for larvae. Artificial saltwater mixes and specific gravity meters (to measure the strength of the seawater) are readily available. The difficulty with water dishes lies in the burrowing habits of most crabs. Many constantly dig up the substrate, so the dish becomes filled with gravel, sand, or dirt every day. Over time this is especially problematic with seawater, since the salt can become concentrated in the soil as the water evaporates.

How Many and With Whom

An important question the keeper must answer when getting new animals is how many can live together safely in an enclosure. Many crabs are gregarious in nature but they can still harm each other within the confines of captivity no matter how well the enclosure is set up. They can become aggressive if there are limited hides or when fighting over mates. Freshly molted animals are more likely to be eaten if tank mates are not well-fed, though attacks after molts can be unrelated to hunger and removed legs may not be eaten. Greater numbers of males in one enclosure can reduce the tendency for aggression. Two fiddler crab males in a cage often ends in the death of the smaller one, whereas half a dozen or more males together can spread out the aggression. (One male at a time may still face the brunt and larger numbers may only slow down the eventual emergence of one surviving male.) When and if it is possible to keep crabs with other types of animals is less of a question. Very few animals survive the depredations of even the most docile crab. With rare exception, crabs are vicious predators.

One of the most important questions in sizing the habitat is the number of crabs being kept in the same enclosure. This relates to how

Terrestrial and aquatic crabs like to hide in dark holes during the day.

the chosen species behaves with its own kind. Relatively docile species can still damage each other if crowded. Many crabs are gregarious and exhibit social behaviors in captivity. Unfortunately, one of those social behaviors is fighting. Vampire crabs can be reared in groups with reasonable safety when they are very small. However, the closer they get to adult size the more likely animals will be maimed or killed. Older immatures—even before they obtain notable color—are sometimes found with missing front claws or legs. Once they obtain adult coloration the males set up territories and may severely damage each other during fights. Except for the purpose of mating, adding more than one crab to a habitat may never be worth it. The size of the cage chosen, the number of hiding spots, and feeding frequency should not have to be closely monitored to limit damage

Most crabs leave artificial sponge alone. This piece of sponge had remained intact for nearly twenty years in various enclosures. The damage was inflicted over a two-week period by a large *Somanniathelphusa*.

Sponge islands do not have to be large or covered in soil. Eventually mosses and algae can obscure the surface.

One end of an old *Geosesarma* enclosure. Three adult female *G. tiomanicum* are located in tunnels beneath the liverworts and mosses. Two of the females are holding fertile eggs. The male is in a tunnel on the island on the other side of the water (not pictured).

or death with individualized caging—it is a far safer solution.

In general, it is a bad idea to keep different genera in the same enclosure since they are more likely to fight. I kept a mangrove crab female for three years and she produced eggs each year, but the male died after six months. I could not locate another male, so the last two batches were infertile. At the end of year three I had a red claw crab male I had been maintaining for six months without finding a female. Both species were previously kept in groups with no aggression, even though there were multiple males in relatively small habitats with a floor space of 20" x 11" (51 cm x 28 cm). I thought it would interesting to see if they might mate, so I put them together. The male showed the female a lot of interest the first day, but I did not read the behavior correctly.

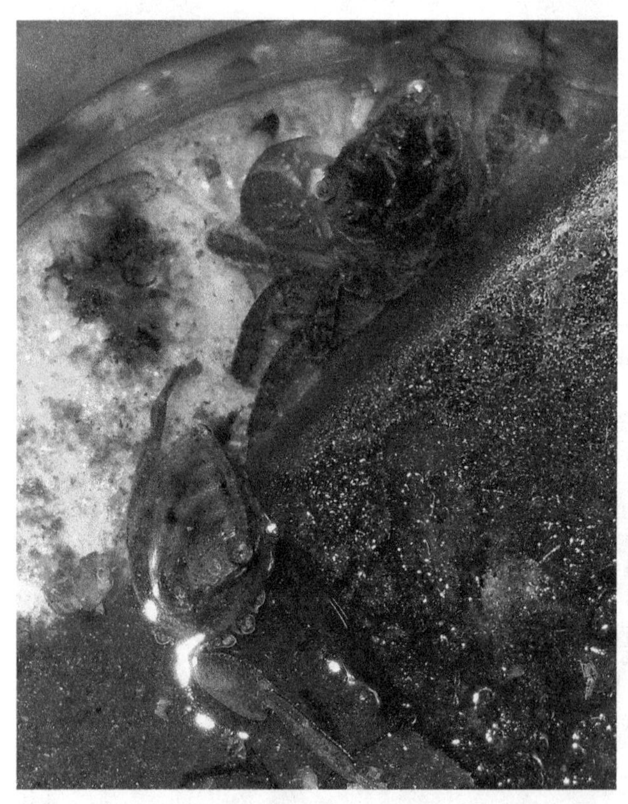

Red claw male and female mangrove crab cohabitation failure, the female was dismembered overnight.

The next morning the poor female mangrove crab had just two remaining rear legs. She was alive, but had little chance of successful regeneration. Closely related species do not necessarily have a similar negative reaction. I had three adult female *Geosesarma dennerle* and a single male *Geosesarma hagen*, with no easy way to acquire more specimens, so I placed them together to see if they would get along or mate. I checked on them every few minutes, then hours, then days. I was pleasantly surprised to see no aggression (though no mating nor reproduction was observed). Terrestrial hermits, all from the same genus, also do not seem to display aggression between species. I have kept many different species of marine hermits together in my reef tank without aggression between similar genera, though big hermits and arrow crabs will eat the small hermit species.

The decision to keep crabs together in enclosures is often practical, but more often emotional. When raising large numbers of immatures, there is rarely enough time in the day to feed, monitor, or clean many dozens or hundreds of different cages. (Survival at first may even be better in groups since only one enclosure needs attention.) Keeping multiple specimens in a single enclosure is more often associated with the human desire for company. There is little evidence that crabs experience the same desire, while refusal to molt in groups and fighting are evidence to the contrary.

Food

Crabs are familiarly known as omnivores, but most species tend towards the carnivorous side of the spectrum. Diets are species-specific and can vary surprisingly for animals with similar-looking claws and body shapes. Even the vegetarian crabs will scavenge dead animals. Most species feast on fish, other arthropods, slugs, worms, and bivalves. Crabs are scavengers and

This strawberry hermit crab (*Coenobita perlatus*)
was just feeding on this fresh strawberry.

Gecarcinus feeding on a chunk of fresh coconut

Coenobita are often immediately interested in fruit like this chunk of watermelon.

will usually take dead prey as well as live. The tiny pointed claws seem hardly capable of grabbing fast-moving prey, since it is like catching a fly with chopsticks. However, some crabs, including *Geosesarma, Metasesarma, and Rhithropanopeus*, can catch fruit flies and crickets on land or daphnia and mosquito larvae from the water with surprising dexterity. Some marine crabs have small disks at the end of their pincers which are designed to aid in removing algae from the surface of rocks. Many crabs have one enlarged claw outfitted with massive molars to crush hard-shelled prey like snails and mussels.

Live prey offers good results for many crabs. Living crickets, mealworms, ghost shrimp, tubifex worms, fruit flies, and brine shrimp are commonly available from local pet shops. Live prey such as isopods, springtails, snails, and daphnia can be collected or cultured. Culture starts can be bought online. Most of these and more can be bought frozen or dried, but they do not elicit the same feeding response and are more likely to foul the cage.

Nearly every species will eat pelleted fish food or hermit crab food. The pellets are most often composed primarily of fish meal. Varied sizes and brands of pellets can be used for different-sized crabs. Fish flakes are often made of the same thing as pellets, but they decompose faster, sometimes in less than a day, are difficult for large crabs to eat, and produce a lot of uneaten waste. Strips of processed fish (imitation crab) are useful foods for most species and they do not decompose as quickly as flake or pellet food. Canned mussels can also be used. Strips of cooked chicken, ground beef, or bacon bits can be used, but lead to a greasy water surface, terribly fouled water, and death, unless used with great reserve. Raw fish and other raw meats are no more attractive to the crab and can be dangerous if you touch your eye or mouth (salmonella, *Vibrio vulnificus*, and other bacteria). Raw foods are also

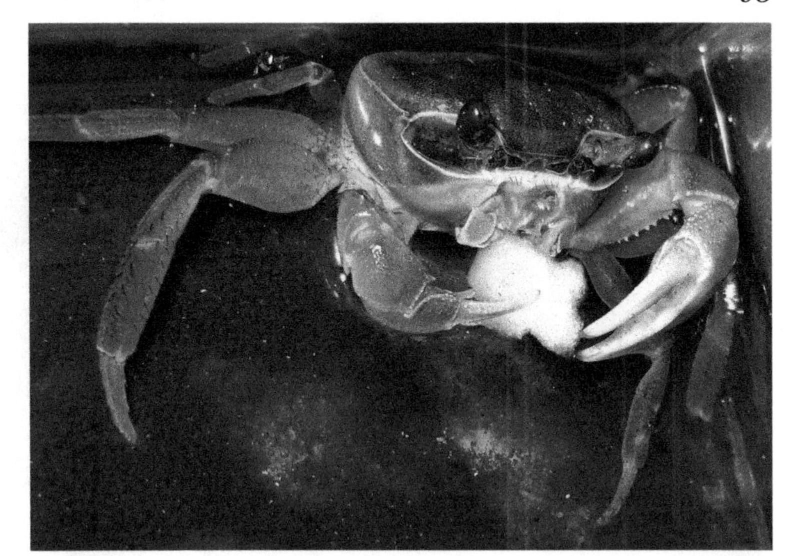

Cardisoma armatum eating a chunk of fresh apple

Aquatic plants can harbor a variety of aquatic creatures. A large number of these swimming mites emerged from a handful of washed up plants on a Lake Erie shore. These appeared to be harmless. (The mites were eventually all eaten by *Hydra* that came in on the same plants.)

Living black worms are commonly sold at pet shops to feed fish. Most crabs will eat these worms.

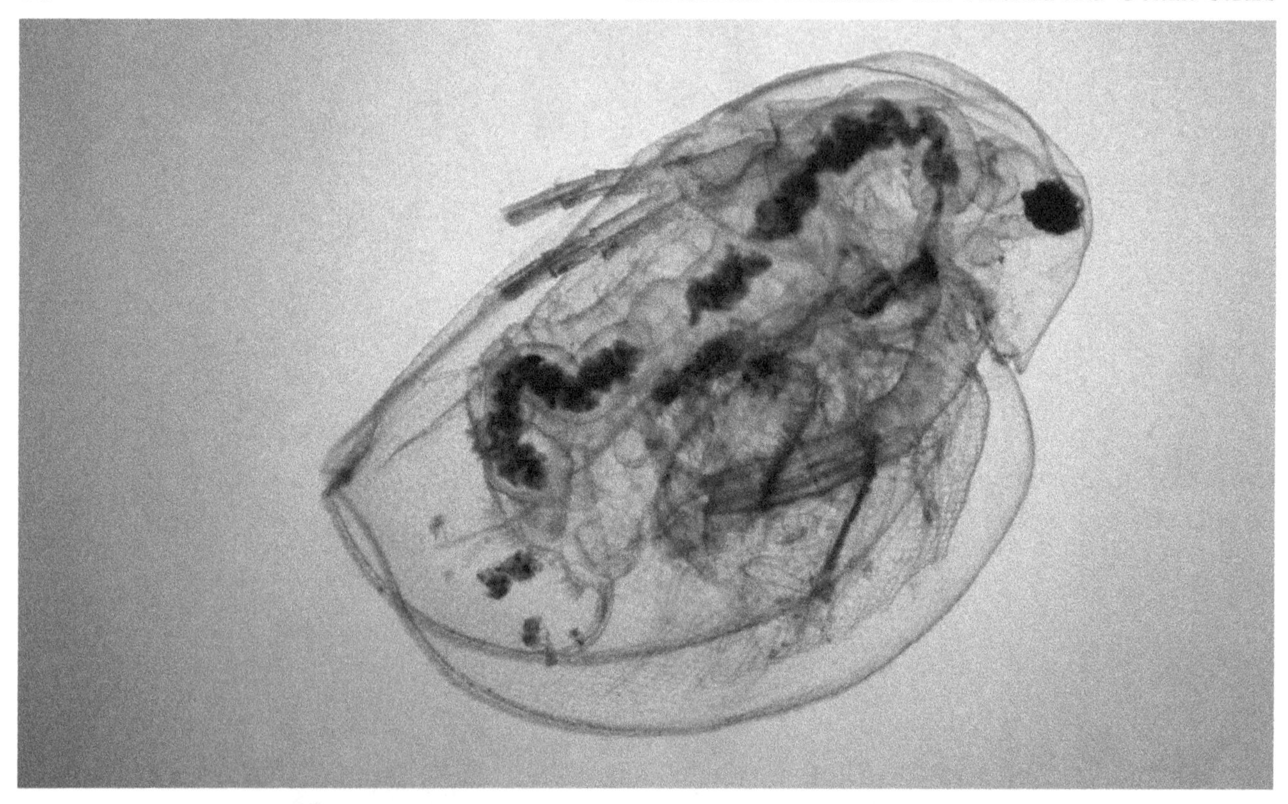

Daphnia magna up close. These freshwater crustaceans are fantastic live prey for small and large vampire crabs.

Some crabs enjoy eating other crabs.

Minor and major *Uca pugilator* feeding greedily on imitation crab.

Appropriately sized crickets are excellent food for *Geosesarma* and are often caught as soon as they hit the substrate.

Hair algae is usually ignored by vampire crabs, but this fresh piece is getting some attention.

Geosesarma eating an earthworm.

more difficult to break into appropriately-sized pieces. Portions of animal matter food should always be sized so the crab can eat the whole thing within a few hours. Frozen small shrimp are great for large species (cooked and peeled, usually sold as "salad shrimp"). Canned small shrimp are less expensive, but they crumble into pieces that are difficult to remove if uneaten. Dried krill is commonly available, but very few species eat it right away, so it begins to decompose after a few days. Excess food of any type should be removed if not eaten in twenty-four hours. Overfeeding in water leads to high levels of ammonia and nitrite that can permanently damage the crab's gills and make breathing impossible.

Common plant matter foods can be offered to crabs, including fruits and vegetables, dead leaves, aquatic plants, and algae. Slices of apple, pear, peach, melon, cantaloupe, grape, carrot, and coconut are eaten by many crabs. Pineapple, banana, and mango are often ignored. Some seeds and nuts including sunflower seeds, almonds, and walnuts are high in copper, so I have not tested them. (Too much copper is harmful for invertebrates because their blood is copper-based, just as humans overdose easily on iron.) Dried, brown, dead leaf litter can be collected from a pesticide-free area. Otherwise, almond leaves or others can be bought by mail order. Some crab species (*Perisesarma bidens* and *Gecarcinus quadratus*) tear through dead leaves, while others wholly ignore dead leaves or eat them sparingly. Aquatic plants can be harvested from a local body of water, though parasites might also be collected. Dried seaweed (algae) is available for sale at some markets or from any pet store with marine aquarium supplies. Plant matter usually takes much longer to spoil than meat-based foods and the spoilage is usually not as harmful. This depends on the plant matter—carrots, aquatic plants, and dried leaves almost never lead to spoilage capable of harming a crab. Sugary fruits are not so benign. Though spoilage is not as rapid or as common as it is with

Common aquarium snails and *Daphnia magna* are attractive to different predatory crabs.

Orchid vampire eating an isopod

Red apple crab eating a mayfly

Vampire crab eating a mealworm

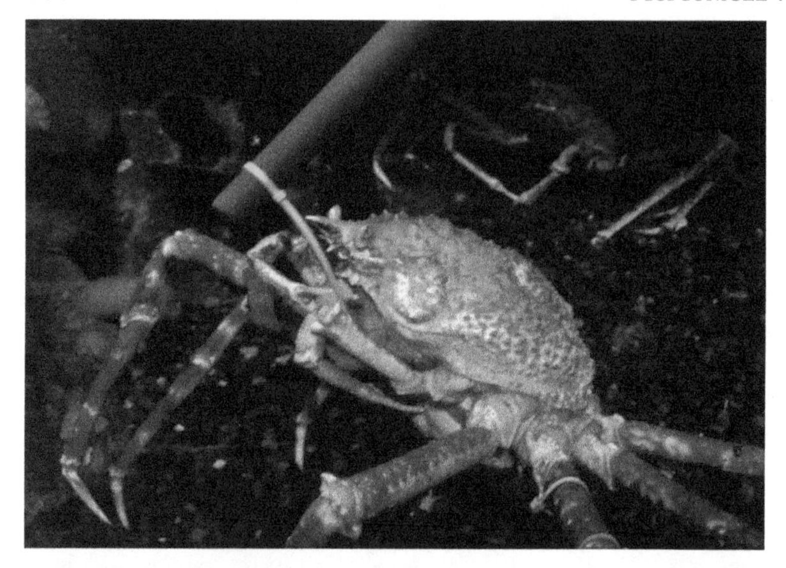

Japanese giant spider crab eating a gigantic jumbo shrimp.

Emerita mole crabs extend filter-feeding antennae to strain the water for plankton.

Antennae are moved up and down and later withdrawn and scraped clean by the mouthparts.

Red claw crab feeding on a piece of imitation crab (processed fish)

Frozen salad shimp and processed fish are great fare for most any crab.

Red claw crab eating dried hardwood leaves

spoiled meat, sugary fruits can also decompose and lead to ammonia and nitrite spikes in the water. I had a piece of spoiled peach lead to the death of two mature dwarf mud crabs.

There are three main types of overfeeding—the first is when food is not eaten and allowed to build up into decomposing piles; second, is when scraps or uneaten portions decompose; third, is a voracious species eating huge quantities of food while the filtration is not keeping up with the crab's excretion. The way to avoid the first problem is to pay attention and not feed crabs food they show no interest in. It is not always easy since food can find its way under decorations or become buried in the substrate. The second is similar, but can usually be avoided by use of clean-up crews, because the leftovers are in small quantities. The last is to remember crabs do not need a lot of food, and a small portion once a day is plenty even for a very hungry crab. A larger or better filtration system and regular cleaning also helps. Overfeeding can be dangerous on land as much as in water. Rotting food on land can poison the air and lead to grain mite outbreaks. The grain mites have a traveling stage and stick to

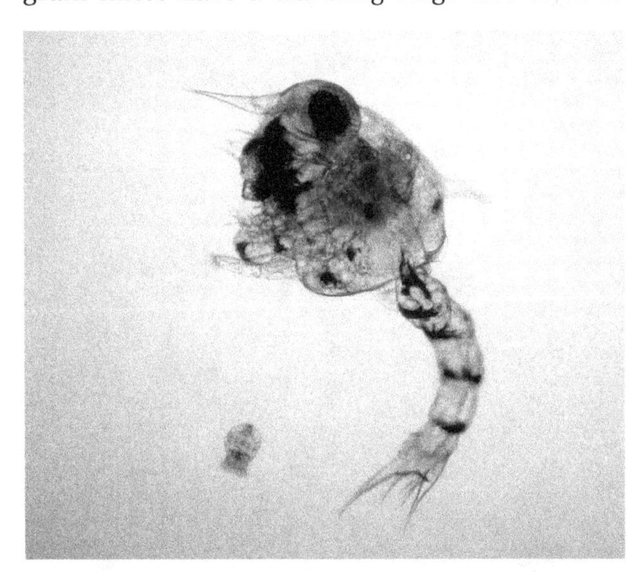

An eleven-day-old *Uca pugilator* next to a rotifer that is a good size to eat.

Moon crab eating a dried date

Moon crab eating a chunk of apple

Moon crab eating a thawed shrimp

Patriot crab eating a large, waffle-sliced carrot

Large immature *Geosesarma* feeding on pieces of processed fish

Uca minax eating processed fish

invertebrate exoskeletons. In high numbers they can restrict or block air passages. The size of the enclosure and filter is important to the volume of food that can be wasted without spoiling the habitat, but there is always a limit.

In freshwater aquaria, small ostracods (seed shrimp) make fantastic clean-up crews. Freshwater daphnia and copepod cultures may include seed shrimp, but neither are useful cleaners themselves. In saltwater, copepods and gammarus (scuds) can be useful in reducing leftovers. Terrestrial clean-up crews are useful for reducing small amounts of leftovers. Common clean-up crews for terraria include springtails and micro-isopods, both of which are commonly available from vendors. Larger isopods work better, but most are intolerant of extremely soggy conditions and crabs eat larger isopod cleaners. If the substrate is very loose and the crab builds burrows constantly, the clean-up crews may die out from being repeatedly buried.

Handling

Hermit crabs are commonly kept and handled by children. They usually stay in their shell if they are being handled carelessly and are relatively harmless to humans. My earliest memory of the purple claw hermit was when I was seven. I knew it would hurt, but I could not help testing out the large purple pincer sticking out of my rather boring pet. I remember the massive purple blister on my finger more than the pain. Hermits often do not try to pinch, but it depends on the crab and the child. Please remember when I say, "handled by children," I mean the average educated, considerate child who is aware that dropping a crab, even inside its shell, can kill it. An aggressive, mean child can kill a hermit within minutes.

Unlike hermit crabs, true crabs are not something a young child should ever handle. Even the careful teenager or adult may harm or be harmed when handling the Brachyura. They are very easy to damage accidentally. Most true crabs are easily startled and jump. If they fall a few feet, even onto carpeting, they may be dead the next morning. Once I dropped an immature *Geosesarma* 4 feet (1.2 m) onto a carpet over cement during a cage transfer—two walking legs fell off overnight. Even if the legs are not missing right away some are likely to fall off by the next morning. Shed legs grow back, but they are usually shorter than the original, may never be 100% even after many molts, and do not always regenerate. Another time an adult female jumped and fell half as far, but she was gravid and died the next day. I have had fiddlers, marsh, and patriot crabs pinch me and leave their claws behind, so I try to avoid handling that is not necessary for cage cleaning or transfer.

The largest crabs (whether terrestrial or marine) can easily inflict injury that requires stitches. The largest species do not have the force to cut through the bone of a pinkie finger, but ligaments, tendons, and muscles are fair game. Small crabs like *Geosesarma* and female *Uca* can have small pincers incapable of causing more than superficial injury, but the crabs still should be left alone because they are easily hurt by a careless person.

Autotomy is the shedding of a body part to facilitate escape and unfortunately many pet crabs use this with reckless abandon (hermits do not). Like autotomized lizard tails, the shed crab leg often writhes or twitches to attract the attention of the predator as the crab makes good its escape. *Sundathelphusa* spp. from the Philippines autotomize the claw while it grabs the predator (or a human finger). The claw continues to enhance its grip by muscle contractions, forcing the perpetrator to concentrate on removing the painful pincer, which gives the crab time to escape (Schwarz, pers. comm.). *Armases* from Pine Island, Florida, also readily give up a cheliped, but mostly in the smallest specimens. The claw grips tightly, but is too small to cause pain. The legs are designed to break off after the coxa (the segment attached to the body) and trying to restrain an active crab during handling will almost certainly cause legs to break. Even if the leg stays attached, it can fall off later that day. Unfortunately, sick crabs may also drop legs and sometimes legs are lost for no apparent reason. Newly imported specimens that molt within the first few months grow back missing legs, but often lose a few in the molt. Fighting among crabs is a common cause of limb loss in captivity.

Molting animals should never be touched. It is very easy to damage or deform the exoskeleton. It is not possible to "help" a molting crab out of its old exoskeleton without damaging it. Molts progress over minutes and if you encounter a crab halfway out, it was probably that way long before you found it. Observe for a movement and if there is no progress in fifteen minutes you can dispose of the animal or find a larger crab in need of a meal. A halfway-molted crab cannot be saved.

Crabs need exercise, but generally they are capable of getting more than enough in a properly-sized enclosure. However, there are "crab balls" available for hermits to roam around the house (like a hamster ball). If the ambient humidity is not too low, and the animal is checked on from time to time, it should not be harmful, though the risk may outweigh a theoretical benefit.

Hermit crab, Howland Island National Wildlife Refuge (USFWS)

Geosesarma sp., Malaysia (Bernard Dupont)

POPULAR SPECIES

The following accounts are meant to provide individual details for common species available to the enthusiast. Family designations are mostly those found in Ng et al. (2008) and are mentioned to show the relationships among commonly kept species. Availability refers primarily through the end of 2017 in the North American and European hobbies. The listed size measurements were taken from actual specimens with Vernier calipers, measuring across the widest upper surface of the carapace. Species may grow larger, but claims of incredible size unsupported by the application of a measuring device are qualified or avoided. Feeding refers to experiences with commonly available food items that result in adequate growth and development; there are certainly many foods that have never been tried. Reproduction includes gender determination, mating, and early development wherever possible. Cohabitation refers to experiences in captive habitats, however a species may react differently when aspects of the artificial enclosure are different—no captive habitat can truly approximate wild conditions.

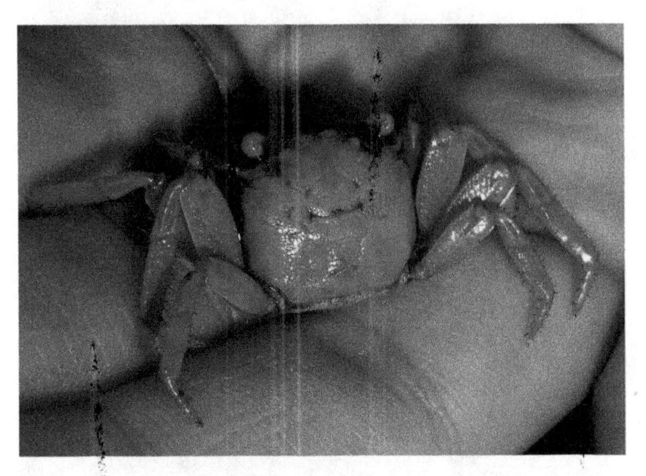

The vampire bat symbol is discernable, even when not obvious, on the carapace of most adult *Geosesarma*.

SELECT BRACHYURAN CRABS: FRESHWATER AND BRACKISH

Vampire Crabs

Geosesarma spp.

These brightly colored land crabs come to the terrarium hobby mostly from Indonesian islands, including Java and Sulawesi. However, collection locations are closely guarded and misinformation seems to be the method used by collectors to protect their hunting spots. The genus ranges widely across tropical Pacific islands, east to Hawaii and west to India. There are currently fifty-three described species, with new species being described on a regular basis. Vampire crabs are colorful, beautiful, and harmless display animals. Some say the vividly colorful eyes are the vampire "fangs," or the markings on the face somehow resemble the face of a vampire.

Geosesarma hagen adult female and male

Geosesarma dennerle female (*left*) and male (*right*)

Nevertheless, the true origin of the common name is a dark marking on the carapace that looks like a bat. The bat shape arises from indentations in the exoskeleton. The indents are usually dark in color and contrast with the light adult coloration of certain species. The following species have been or are still commonly maintained in captivity.

Red Devil
Geosesarma hagen Ng, Schubart & Lukhaup, 2015
Java, Indonesia
Older adults with dark black legs are usually 12-14 mm across the carapace. This species can be sexually mature at 8.8 mm to 11 mm. Small mature animals have banded legs (not solid black) and their carapaces transform from red during the day to brown or orange at night.

Purple Vampire
Geosesarma dennerle Ng, Schubart & Lukhaup, 2015
Java, Indonesia
Mature males of this species are usually visibly larger than females. This species is lankier, with longer legs and a proportionately smaller body than other regularly available species. Males rarely reach 14 mm carapace width and females seldom exceed 12.5 mm.

Mandarin Crab
Geosesarma notophorum Ng. & C. G. S. Tan, 1995
Known from Pulau Lingga, an island east of Sumatra (Indonesia)
Unfortunately, any small orange-clawed sesarmid crab usually ends up being sold as a 'mandarin.' *Geosesarma notophorum* is often touted as the only species known where the early immatures climb onto the female's back to be carried around, similar to young scorpions, opossums, and certain frogs. The species name means "back-bearing" in Greek. However, as our knowledge of reproduction for various species grows, this behavior may prove somewhat common. *Geosesarma nemesis* Ng, 1986 from Singapore is also known to carry young on its back. *Geosesarma pontianak* Ng, 2015 was sold as *G. notophorum* for some time and carried young on the carapace after hatching (Rademacher & Mengedoht 2011).

Tangerine-Head Vampire Crab
Geosesarma krathing Ng and Naiyanetr 1992
Eastern Thailand
Specimens began showing up in 2007 and were being reared in Europe at least as early as 2009 (Dost 2009). Adults rarely exceed 13 mm in carapace width.

Geosesarma malayanum Ng & Lim, 1986
This is a very rare import from Malaysia whose only common name is also mandarin crab. This species is famously known to be found in *Nepenthes* pitcher plants in nature.

Geosesarma bicolor Ng & Davie, 1995
Java, Indonesia
This crab has purple legs, red claws, a purple and gray body, and yellow eyes. Many of the commonly kept species or color forms (specifically *G. dennerle* and *Geosesarma* sp. "orchid") have been labeled bi-color vampire crabs, but none match the described color for this species (male gonopods would be needed for identification).

Disco Vampire
Geosesarma tiomanicum Ng, 1986
This beautiful pastel-colored species (AKA sandfire or "aristocratensis") is of uncertain origin—it comes in with Indonesian shipments supposedly from Sulawesi. The hot pink coloration of the adult takes almost two months to develop after the ultimate molt. Wild adults can be up to 15 mm

Adult male *Geosesarma tiomanicum*

Young adult pair of captive-reared disco vampire crabs, *Geosesarma tiomanicum*

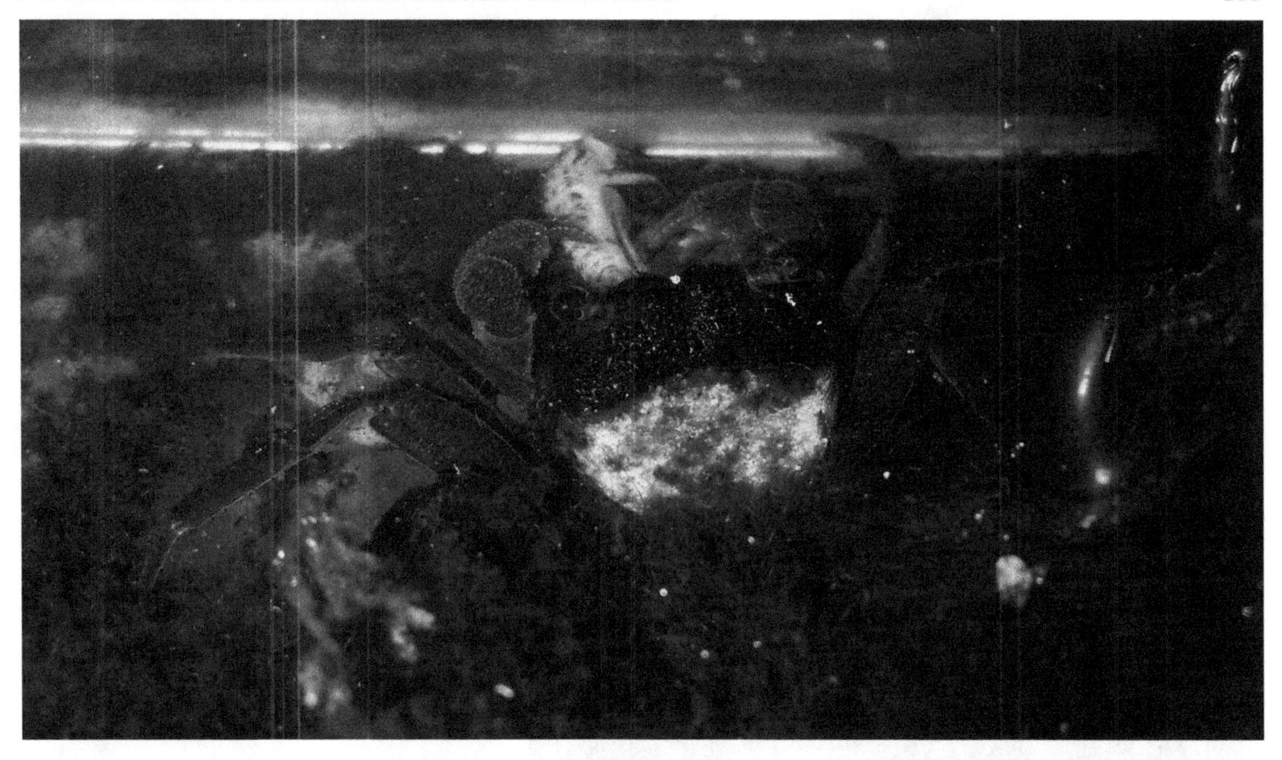

Geosesarma dennerle catch crickets on land and may bring them underwater to feed

Geosesarma dennerle female with a mostly yellow carapace

Geosesarma dennerle: vampire crab females can be somewhat cryptic as adults.

Red devil crabs (*Geosesarma hagen*)

Geosesarma tiomanicum captive-bred young adults discovered mating beneath a piece of bark

Orchid vampire exuvium abandoned in the water

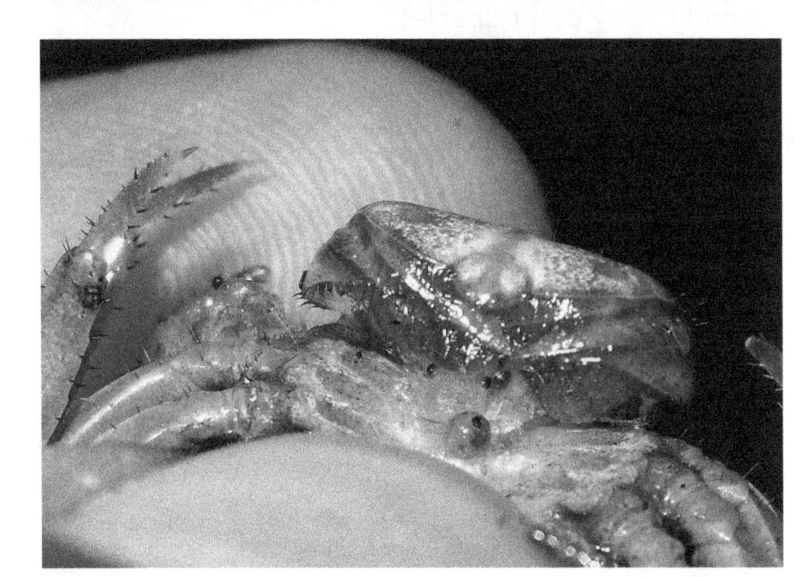

Geosesarma tiomanicum egg and hatchlings

Undescribed *Geosesarma* sp. from Panay, Philippines. © Christian Schwarz

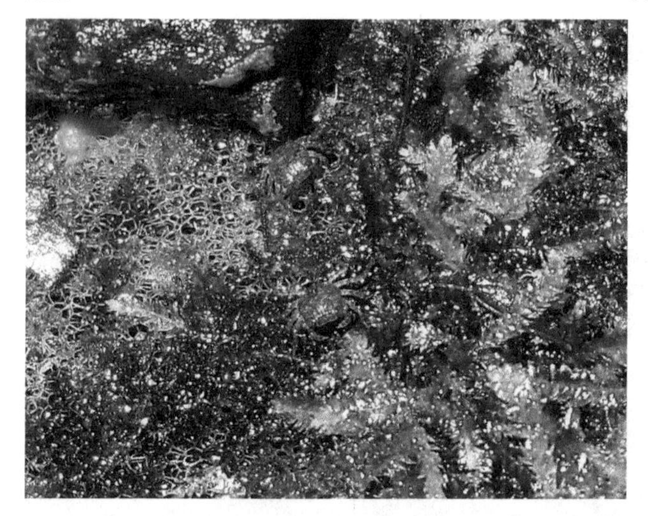

Orchid vampire early instars are brown and resemble any other species.

across the carapace. Captive adults have obtained sexual maturity (initial) as early as 11 mm and 225 days (McMonigle 2017b).

Orchid Vampire

Geosesarma sp.

This was probably the first vampire crab and it was available with regularity from 2006 to 2011. It was called the Celebes jungle crab before the vampire name stuck. Captive specimens were available long after imports stopped, at least through 2015. This species is larger bodied than *G. dennerle* or *G. hagen,* usually reaching 15 mm carapace width. Adults of both genders are approximately the same size (females are just slightly larger on average).

Captive-reared adults usually live two to two-and-a-half years after reaching maturity. The rare specimen has lived four years in adulthood (pers. exp. *Geosesarma* "orchid"). Captive-reared adults do not die early without obvious cause. Wild adults often die within six months for no apparent reason and rarely live more than a year. This may relate to natural age, shipping and handling practices when removed from the wild, or wild stresses such as parasites. Wild specimens are often weakened and some will die within days or weeks of arrival. Imported specimens are also likely to drop a leg or two in shipping or within a few weeks of arrival. Wild-collected adults seem to fight more, but it may not be possible to differentiate stress autotomy from the results of aggression. Wild adults may live as long as captive-bred, it this is just not the normal result. One pair of wild-caught *Geosesarma hagen* was still alive after 3.5 years but never produced young (pers. comm. Joe Reich).

Geosesarma hagen, 3rd instar

2nd generation orchid vampire, 2012

A male *Geosesarma* sp. in Panay, Philippines © Christian Schwarz

Undescribed *Geosesarma* sp. from the Panay rainforest, Philippines. All species have nearly identical indents in the carapace, but the adult coloration of some species does not include a darkened bat marking in the middle. © Christian Schwarz

All the *Geosesarma* species have very colorful adults and coloration may be used to identify species that have unique coloration. On the other hand, immatures of various species look pretty much identical. Since the mature male genitalia are needed for identification, it is unlikely any taxonomist could identify an immature vampire to its species. Small crabs are a cryptic, mottled mixture of brown and they are able to change color to match the darkness of the habitat's background. As they grow closer to maturity some of the adult coloration begins to show. Immature *G. hagen* start to show a little orange a few molts before maturity, while most show some purple and may not be able to be differentiated before achieving full adult coloration. Captive-bred specimens are just as brightly colored at maturity as wild-caught crabs are, however they can reach sexual maturity before obtaining full adult coloration. Some coloration, specifically the hot pink of the adult *G. tiomanicum* carapace and legs, takes two months to develop after molting to maturity.

Availability: At least two or three species are available through mail order year-round. *Geosesarma* even show up at diversified local pet stores on occasion.

Adult Size: Carapace width at maturity is 8.5 mm to 15 mm.

Specific Housing: For the half dozen or so commonly available species, the enclosure requirements seem to be uniform. These small crabs are easy to house since they drink only fresh water, and do not require a clean, well-oxygenated source. Adults can survive for months or longer with a few millimeters water depth. A two-and-a-half-gallon to ten-gallon (9-38 liter) glass terrarium makes a good enclosure. I prefer not to use lids because stagnant, high moisture can kill them within days. A covered, decorated terrarium with small fountains or streams (powered by small water pumps or air pumps) and open space is an alternative habitat design to keep the oxygen levels adequate. They are not very destructive and rarely eat plant materials other than certain cut fruit and vegetables. They are predatory and cannot be housed with other small creatures—even tiny springtails can be caught by adults from time to time (while small crabs eat springtails with reckless abandon). Hiding areas or deep substrate should be avoided or the crabs will remain permanently out of view. Too many hiding areas make it difficult to gauge if they are feeding, need food, have had young, have molted, or are long dead. The cage should have water and land areas that are easily accessed with a scalable, sloping ramp. They cannot climb smooth surfaces, which is the reason a lid is unnecessary to contain them. The adults can usually access a shallow dish (such as the lid of a 1 oz. cup) but the young are too small. This genus is considered terrestrial and larger specimens are capable of molting on land, but they prefer to molt under water. A paludarium is suggested for these species. One option is to cut pieces of 2″ (5 cm) thick sponge and keep the water level below the top of the sponge. The size and shape of the water areas is easily adjusted to individual preferences. Adults tend to stay on top of the sponge while young crabs are often on the sides just below the water line if there is plant cover. A thin layer of dirt can be placed on the sponge. Various mosses, including aquatic Java moss that can also grow on land, and liverworts can be grown on top to hold back the dirt and provide decoration.

Temperature and humidity requirements are somewhat controversial. 85° F (29° C) is often recommended as optimum, but I have reared numerous adults from hatchlings and successfully produced young in the low 70s without any molting or developmental problems. Development of my first pair of captive-reared *Geosesarma* sp. "orchid" was still rapid at room temperature, less than five months (McMonigle 2011), though

Geosesarma hagen fresh import missing a few legs

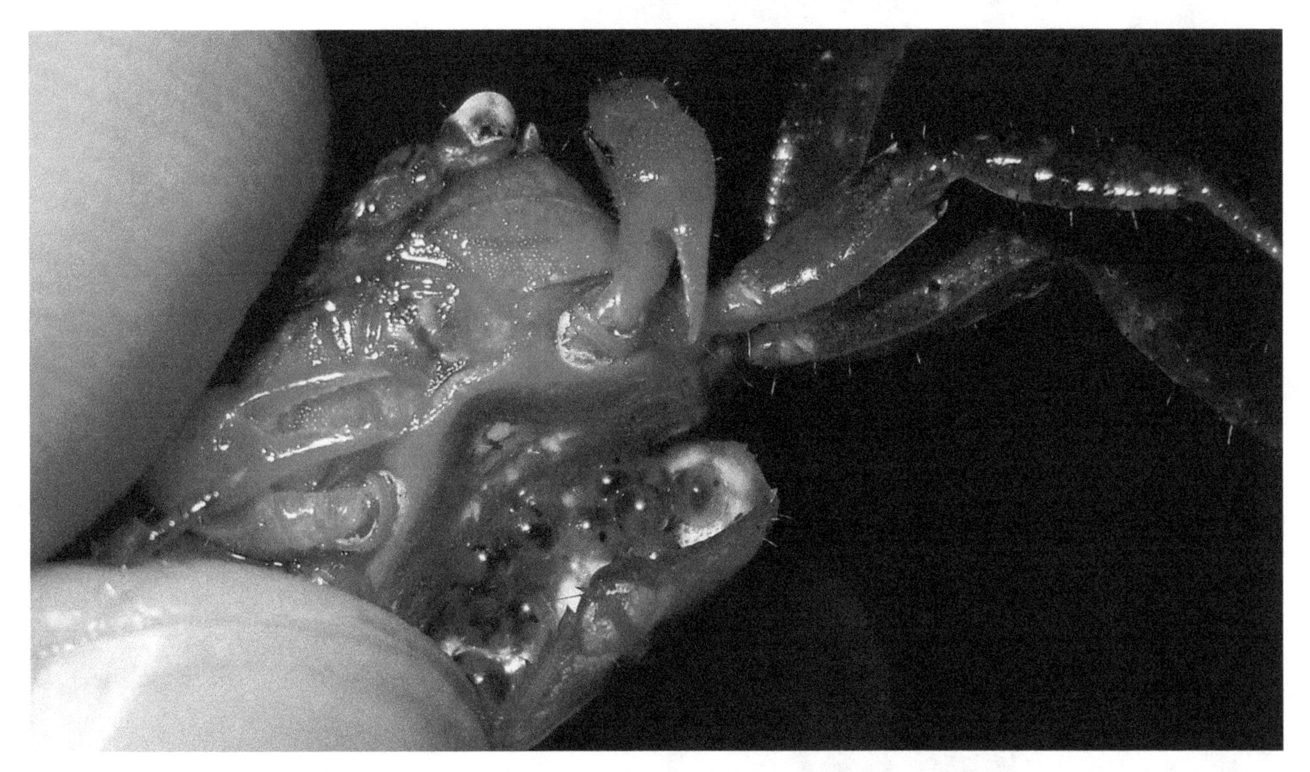

Geosesarma hagen female with developing eggs

This orchid vampire is chewing apart an isopod. This genus is highly attracted to mobile prey.

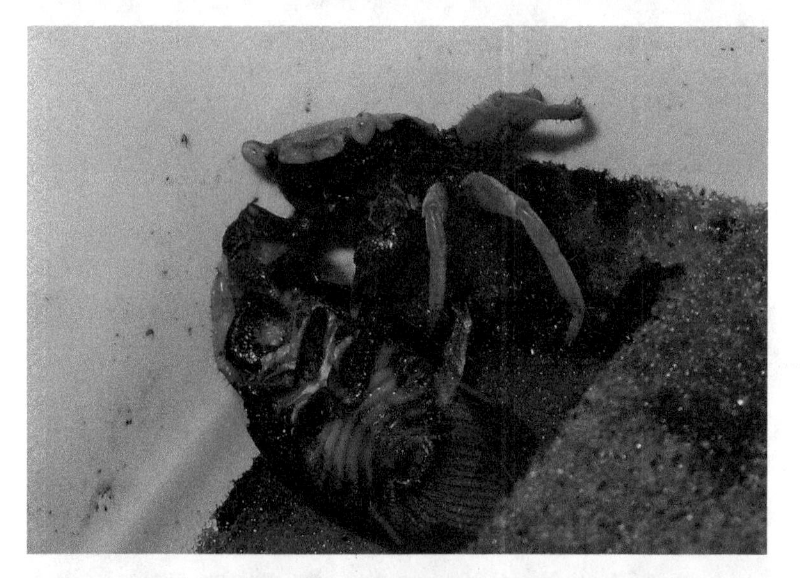

Geosesarma eating a large roach

Geosesarma tiomanicum eating an isopod

they were already the size of a bb pellet when acquired. Later generations took somewhere in the range of ten to twelve months to reach full adult coloration at an average of 72° F (22° C). Disco vampires progressed from dispersal to sexual maturity in 7.5 months at approximately 72° F (22° C), but they mated before obtaining full coloration. A temperature of 68° F (20° C) can result in the death of very young immature crabs over time, though it does not bother large immatures or adults. Much lower temperatures are likely deadly to all stages. High humidity is a common recommendation that caused the death of most of my first *Geosesarma*. A terrarium without adequate air movement can suffocate the crabs within a few days (or less) if the lid has only very tiny air holes. Mechanical movement of air or water within the enclosure should still allow for gas exchange with the ambient atmosphere even if the lid is mostly sealed. I have always used an open-top container since the accidental death of some captive-hatched immatures in 2010 caused by limited ventilation. Measured humidity should be immaterial if there is a water feature in the enclosure.

FEEDING: Any small terrestrial or aquatic creature that moves to attract attention works well (springtails, roaches, isopods, daphnia, ghost shrimp, mosquito larvae, tubifex worms, black worms, etc.). Small crickets and mealworms are a great treat for older specimens. Fruit flies can be caught quickly, but run up the sides so rapidly that most escape. The primary diet does not have to consist of live prey. They can be raised from hatching to adult on fish food pellets (those made primarily from fish and shrimp meal) and small pieces of processed fish (imitation crab). They will feed on a wide variety of animal and vegetable matter, but are primarily predatory and hunt by sight. Small pieces of food tossed in to the cage often elicit a response, but active prey is better. They show limited interest in fruits and vegetables unless very hungry.

REPRODUCTION: In the first few years of looking for *Geosesarma* I would often buy or trade for "pairs" that were actually two males. Adult males and females appear similar in size and general appearance, but they are easily sexed. The most obvious difference is the size of the abdomen (tail) that is held up against the bottom of the carapace. This is where females hold the eggs before hatching and it is very wide. The male's abdomen is narrow and folds into a groove. The difference is extremely easy to see on mature specimens. Smaller specimens can also be sexed by the relative width of the abdomen though the difference is not as pronounced. It is not possible to see any difference on the abdomen width of very young crabs before the carapace width is 3-4 mm. Other sexually dimorphic characteristics include the shape of the front claws and the body coloration. These characteristics cannot be seen (or show up only slightly) before the molt to maturity. The males have much larger, thicker claws when they are fully grown. Males also tend to be more brightly colored, often much more so. If a pair is equally colorful they are very likely the same gender.

Reproduction information is limited to wild-caught specimens and a few dozen captive-reared adults over a few generations, but it seems female *Geosesarma* sp. "Orchid" wait a full year after reaching maturity before producing eggs (Reich 2010, McMonigle 2011). I have observed young captive-reared adult *Geosesarma tiomanicum* mating, which resulted in a gravid female (young dispersed from mother on 9-23-2016, grew quickly to sexual maturity, and mating was documented 7-18-2017, 11 mm carapace width). Two females were holding eggs 11-20-2017. Males can mate with females above water and when females have not undergone a recent molt. Mating is rarely seen, but there is another sign of pairing. Caging should have a number of hides and each crab has its own burrow. If a male and female stay in the same hide for a week or two, the female is often found carrying eggs in the next two months.

The eggs are not visible from the side like they are for many crabs, but there is visible evidence. The tail of the female is raised less than a millimeter—it is not dramatic but it is usually identifiable unless there are very few eggs. The hatchlings are much larger than the eggs, so the tail becomes more elevated a few days or weeks before young are released. Even when there is a large hatched brood, the pleopods cover the edges so eggs and young are almost never visible before they are released.

Geosesarma eggs are large (1.4 millimeters, nearly ten times the diameter of a blue crab egg) and there is no planktonic stage. The immatures are free-living at around 2 mm across the body. Hatchlings can darken or remain white until after the first molt. The hatchlings from *G. tiomanicum* measured 2.1 mm across the carapace. After ten to twenty-one days they molt to approximately 2.8 mm. Another three weeks later they molt to 3.9 mm.

It is important that brooding females are not handled. After being handled, a female may reach in and start flinging out immatures with her claws (McMonigle 2011). A disturbed female can drop undeveloped, healthy eggs if disturbed (McMonigle 2016). The mother will not re-gather eggs or immatures, and discarded young can be weak and experience high die-off depending on how premature the separation. Eggs gestated artificially can survive for weeks, but they do not hatch successfully. However, it is possible to put the eggs back or to move eggs from a deceased adult female to a live one with limited success if the eggs are still healthy (I transferred fourteen developing eggs from a recently deceased *G. hagen* female to a live female. Forty days later three hatchlings were found).

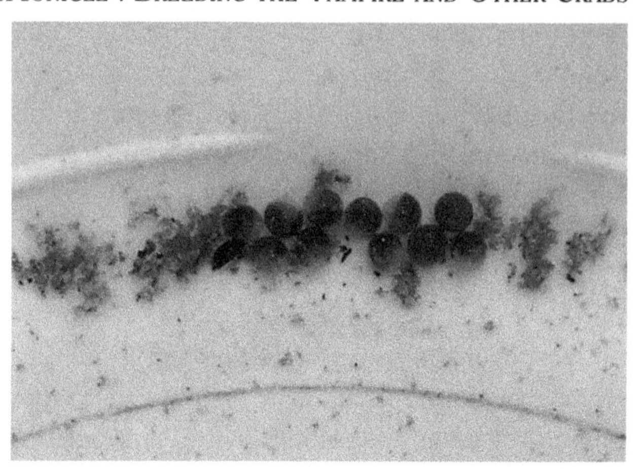

These abandoned *Geosesarma* eggs show signs of development.

Adults eat any immatures they can get their claws on, so young should be moved to their own cage if much of a survival rate is hoped for. Immatures are cryptic and look like little rocks or pieces of dirt. In a heavily decorated tank one or more young may escape detection from the adults and survive to maturity (the keeper also will likely not see them). Young crabs of the same age can be reared together with minimal losses, but as they grow larger, legs will end up missing. This is a sure sign of overcrowding. They molt every few weeks and even stragglers can mature in six months.

Rearing, gestation, and breeding is successful between 72° F and 76° F (22-24° C). I do not have a room that is 80-85 ° F (27-29° C). I tried keeping *Geosesarma* at 80° F (27° C) when I first acquired tiny, captive-bred "orchids" in 2010. Sadly, I ended up killing four out of six in the first two days because I had limited ventilation and maintained the temperature with a small electric water heater. Reich (2010) states they cannot molt correctly in cold temperatures, but there must be another factor, since I have kept the same species as cold as 67° F (20° C) without any molting problems. However, if temperatures stay below 70° F (21° C) for weeks, some specimens

Red apple crab and a fresh exuvium from a different animal

die, especially those in the first few instars. Development from being released by the mother to sexual maturity usually takes nine to twelve months.

COHABITATION: There should only be one adult male per cage. Adult specimens are often kept in groups, but males are known to fight and will tear each other's legs off. They usually fight to the point where the loser has too few legs to recover. Males may get along for a while without incident, but eventually there will be a violent confrontation. Establishing isolated territories in the enclosure using water and decorations to keep males separate only delays the inevitable. One will eventually enter the others' territory and kill or be killed. Across multiple species and hundreds of cohabited adult specimens, I have never seen a male kill a female. However, in a single wild-caught pair of *G. hagen* the female ended up losing both front claws and died after a few months.

Red Apple Crab

Metasesarma aubryi (H. Milne Edwards, 1869)

The red apple crab is fast moving and a good pincher. This species is unusual for a sesarmid because the adults seem to avoid being wet (though specimens are capable of surviving submerged in water for hours). If there is one little island in the middle of a cage, they will stand on it all day long. Adults must molt and stay above water for some time afterwards or they drown. In more complex caging the crabs make tunnels under pieces of wood or stone and will not be seen

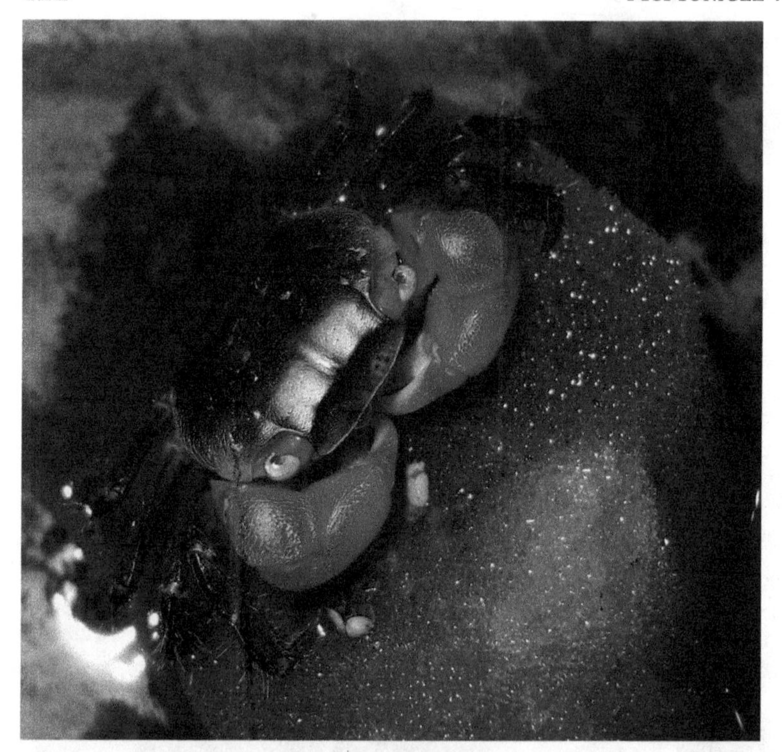

A red apple crab (*Metasesarma aubryi*) mature male with namesake coloration

Red apple (*Metasesarma aubryi*) adult pair, female with smaller claws

Mature red apple crab females can look very different from males.

during the day. Larger plants like bromeliads and ferns are usually left alone, but mosses and liverworts are trampled and torn apart over time (possibly eaten). This species is found on shores and islands across the Indo-Pacific from Sumatra, Indonesia, to New Caledonia. It can be maintained in fresh or brackish water enclosures.

AVAILABILITY: The red apple crab has been an uncommon import, though many crab vendors have had some at one time or another since 2008 or so. It comes in with Indonesian imports and the exact location of origin, as usual, is a secret.

SIZE: Specimens tend to be bigger than the larger *Geosesarma,* but not by leaps or bounds (around 20 mm across the carapace).

FEEDING: Pellets, fish flakes, processed fish, crickets, isopods, mealworms, etc. Like *Geosesarma* they will eat certain vegetable matter, but are visibly predatory in nature. They are highly attracted to the movement of small creatures—primarily movement above the water line. If a small piece of fish is thrown at them, they often pounce on it immediately.

REPRODUCTION: Adult males are black with a red margin, while females have enlarged, lighter-colored markings that range from red to light orange. However, they change colors at night and sometimes during the day. On rare occasion a female can be colored the same as the male, but her claws will always be smaller. Her wide abdomen is a dead give-away. Larvae are planktonic and probably have never been reared in captivity. The eggs are tiny, but much larger than *Uca* spp.

COHABITATION: They seem gregarious, especially male and female pairs, and they do not eat molting animals of the same species and similar size. However, newly molted animals may end up with a few legs missing if there is not adequate space. If emergent surfaces are inadequate, they may chase freshly molted animals into the water where they drown. Solitary caging is suggested.

Red claw crab (*Pseudosesarma moeschi*), adult male

Red claw crab males tend to be gregarious and rarely, if ever, harm each other unrelated to a molt.

Red Claw Crab

Pseudosesarma moeschi De Mann, 1888

This species is commonly sold under the name *Sesarma* (or *Perisesarma*) *bidens*. It is sometimes sold as the Thai crab, though specimens seem to be imported through Indonesia or the Philippines rather than Thailand.

The red claw is relatively easy to handle and slow moving. Even the large-clawed males are reluctant to pinch when handled (of course they can pinch). This species is sold as being capable of living in fresh water (not entirely impossible if the fresh water has extremely high hardness and alkalinity, as in limestone aquifer wells), but it should be provided brackish or salt water for survival. The first time I attempted to keep this species the adults began to die after a few weeks for no obvious reason. As soon as I replaced the fresh water with brackish, they stopped dying.

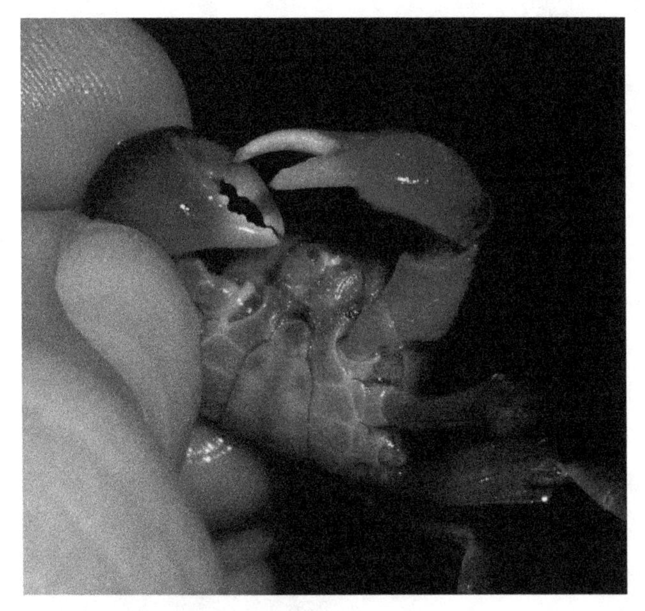

Red claw crab, male abdomen

Red claw crab, female abdomen

AVAILABILITY: The red claw is a common import available from mail-order crab vendors year-round, but it is rarely seen in North American pet shops. Nevertheless, a single large pet chain carries this species with some regularity. The red claw is one of the most common pet species of true crab and though it must be imported from the Indo-Pacific, it is often as inexpensive as native North American fiddler crabs. Werner (2003) calls this, "The species most frequently offered for sale" (in Germany).

SIZE: 16-18 mm across the carapace.

FEEDING: These crabs are heavy feeders and eat almost anything. Fish food pellets, dead brown leaves, fruits, and vegetables are common fare.

REPRODUCTION: I ordered ten through mail order and requested an even gender split, but received ten males. Still, they can be purchased in pairs from some sources. Adult males have a narrow abdomen, much larger claws than the females, and are more brightly colored. The eggs hatch after three weeks. The planktonic larvae are capable of eating baby brine after the first molt and there is photographic documentation of rearing in captivity (Järvi 2009).

COHABITATION: Adult males are normally not aggressive towards adult males of the same species. However, males commonly tear front claws or legs off of females when there is not enough space or too few hides.

Mangrove crab female, 2013

Same mangrove crab female in 2016
(Her exoskeleton darkened significantly over time in a captive enclosure.)

Mangrove Crab

Perisesarma sp.

The common name "mangrove" is almost as vague as "brown" or "red" crab because members of the Family Sesarmidae (including the popular red claw and red apple crabs) and Family Grapsidae ("Sally lightfoot" crabs), among others, are often called mangrove crabs. *Perisesarma* species have wide, flattened legs that look different from the previously discussed sesarmids. The adults are mostly reddish with some red or orange highlights, but the red highlights can be white on different specimens from the same import. Despite the specific uncertainty, in the northeastern U.S. the name "mangrove crab" long referred to the species pictured here. It has been seen in pet stores for more than a decade and reportedly comes from Indonesia. Unlike the red claw crab, this species will molt, produce eggs, and live for years when kept in freshwater. In captivity, wild adult specimens usually live two or three years with little attention required. Full-grown specimens only molt once a year.

2017 Indonesian imports labeled as mangrove crabs have contained an additional species. The primary species has similar-shaped, flattened claws for both sexes, and is colored dark gray overall, usually with red-tipped chelipeds. The second species is tan with bulbous, white-tipped claws. The males of the tan species have chelipeds similar in shape to *Pseudosesarma*, but there are notable tubercles on the upper surface of the cheliped finger as usual for *Perisesarma*.

Availability: *Perisesarma* spp. have shown up at North American pet stores up to a few times a year between 2005 and 2017.

Size: 16-18 mm across the carapace.

Feeding: Pellets, fish flakes, processed fish, algae, leaves, fruits, and vegetables. They are aggressive generalist feeders. This species is very destructive to all types of aquatic plants.

Reproduction: The mature male and female have front claws that are similar in shape. The chelipeds are similar in size for both genders, but only on smaller, younger males. Older males can be picked out by the larger claws. Sometimes the males have red- or orange-tipped claws and the females have white tipped claws; sometimes the colors are reversed. Males and females tend to be around the same body size. Mature specimens are easily sexed by comparing the shape of the tail (the female's tail is characteristically wide for holding thousands of tiny eggs). Females produce eggs about once a year in captivity and the eggs hatch after three weeks. The female can be moved to full salt (1.022 specific gravity) for this period. A few thousand larvae hatch out approximately the same size as freshly hatched brine shrimp. They are somewhat jumpy like baby brine and pale orange in color. Hatchlings die *en masse* on day five or six without adequate planktonic food.

Cohabitation: Adults do not harm each other even during molting. I have kept males and females in groups and have not encountered missing legs or other signs of aggression.

Squareback Marsh Crab

Armases cinereum Bosc, 1802

Armases are semi-terrestrial and need emergent areas to survive. They can be collected near the high tide mark on sandy soil in a range of areas from full salt to mostly fresh water. Marshes and coastal forests are common habitats, but this species accepts a wide range of conditions. In captivity, marsh crabs can move up the vertical sides of smooth-walled containers by climbing on each other. Fortunately, they cannot use this skill to reach more than two or three times their legspan. Small specimens tend to have a monochrome carapace while adults are contrastingly mottled in a checkered pattern. At all ages they have banded legs and are well-camouflaged

Mangrove crab female with fertile eggs a few days from hatching, October 2013

Mangrove crab, the same female carrying fresh, infertile eggs in May 2016

Recently imported crab that arrived with mangrove crabs from Indonesia

An egg-laiden, terrestrial sesarmid from Thailand © Matthijs Kuijpers

Small adult *Armases* with patterned carapace

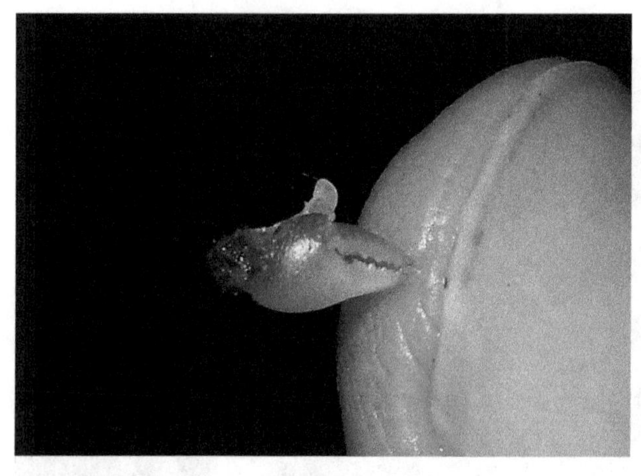

Armases' common defense is to autotomize a cheliped, which continues to pinch for a minute or two.

among dead leaves and sandy soil. *Armases* occur in every shade of brown—sometimes quite a few shades are represented on a single animal. Each animal also changes appearance from night to day, attempting to match the general pattern of a given substrate. When crypsis fails, its next line of defense is autotomy. This can be rather frustrating for the keeper since appendage loss is a key failure in care. During handling, specimens should be grabbed from the rear because one front claw is the sacrifice of choice. (I am aware of no other crab so disposed to ejecting a cheliped).

Availability: *Armases cinereum* can be collected as far north as Maryland in the United States. It is widespread along the Atlantic coast, south all the way to Brazil. This crab is not found in pet shops, nor through mail-order marine vendors, probably because it would rapidly escape any marine aquarium or drown. However, at least one specialty marine collector lists this species as available.

Size: Mature specimens range from 14-20 mm across the carapace.

Armases, female abdomen

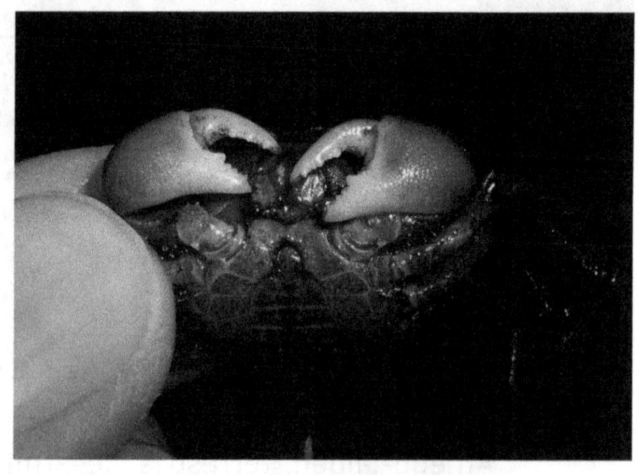

Armases, male abdomen

Armases adult, Pine Island Florida, found at night on land among leaf litter

Armases immatures do not have the dark and light checkered carapace.

Feeding: In the wild they forage in leaf litter. In captivity a wide variety of foods are consumed, including fish food pellets, krill, shrimp, fish, certain fruits, carrots, and dead leaves. Live feeder invertebrates such as isopods and small crickets are eaten. Like most sesarmids they are attracted to movement and investigate objects thrown in the cage. They may abandon items that do not continue to move.

Reproduction: Males and females look similar from above, but the male's chelipeds are visibly larger, even on small adults. The male raises his enlarged chelipeds forward in a threat display. Females produce thousands of eggs in one brood that hatch into planktonic larvae.

Cohabitation: Groups of males and females coexist, at least for short periods, without incident.

Aratus pisonii habitat, mangrove flat in Cape Coral Florida. Adult specimens are seen about two meters off the ground. © Eric Russell

Mangrove Tree Crab

Aratus pisonii (H. Milne-Edwards, 1837)

There is only one species in the genus, which is an amazing tree climber. My initial enclosure was poorly conceived, but fortunately I returned after twenty minutes to check on them. I found one five feet off the ground on the side of a shelf and another a little higher, hanging onto some cords. In nature these specimens were found on 2-4″ (5-10 cm) mangrove trunks about six feet (two meters) off the ground. They jump if disturbed, to disappear into the water or soft mud below. This behavior can be deadly during handling if they land on a cement floor from a good height. *Aratus* is found in tropical mangrove swamps throughout the New World. It occurs along both the Pacific and Atlantic coasts of Central and South America. It is also common throughout the Caribbean as far north as Florida. Males can take up to five years to grow to full size (Warner 1967).

Availability: I have never seen this species on any availability list. It is a common inhabitant of vast tropical New World mangrove forests, but it takes hours for a collector to catch a few, so it is unlikely to make it to vendor lists.

Size: Mature specimens are larger than any of the common imports, 24 mm carapace width, and nearly the same length.

Feeding: In nature this species feeds mostly on mangrove leaves, though it likely hunts small invertebrates among the mangrove roots. In captivity, *A. pisonii* hunts down and eats adult feeder crickets. It also feeds on dead hardwood leaves, carrots, apple, thawed shrimp, and processed fish.

Reproduction: The male has larger, more hairy claws, and is slightly more contrasting in color. The male's abdomen is unusual because it is wide like a female's abdomen, except for the very narrow terminal segment. The concave abdomen hugs small mangrove trunks so the male's unusual abdominal shape may be an adaptation for climbing. Females produce large numbers of planktonic larvae.

Cohabitation: A male can be safely kept with multiple females. I have not had the opportunity to test cohabitation during molting or with multiple adult males.

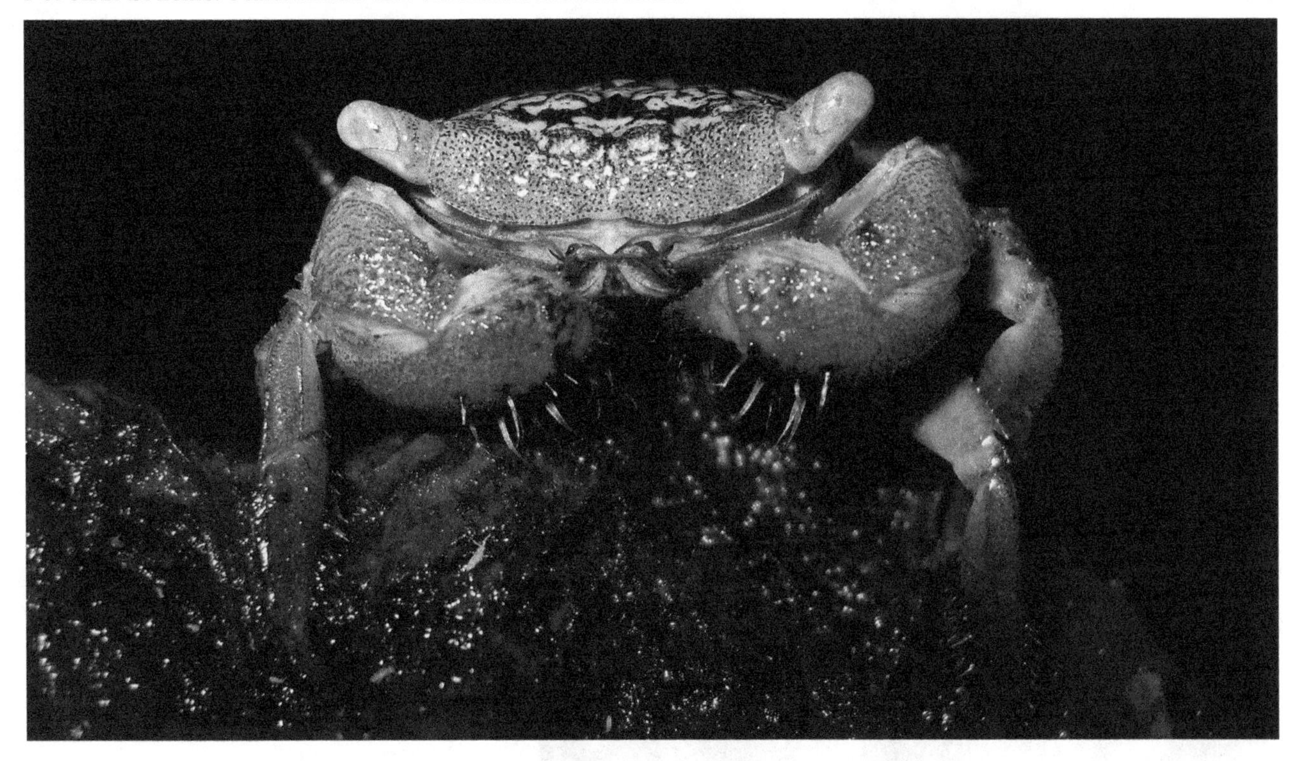

Aratus pisonii male, Cape Coral Florida

Aratus in habitat, Pine Island, Florida © Eric Russell

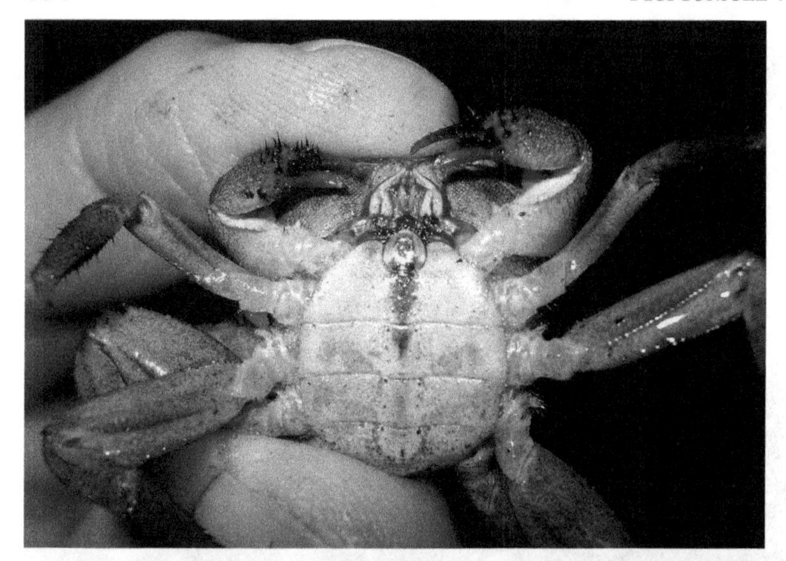

Aratus pisonii, female abdomen

Aratus pisonii, male abdomen

Aratus pisonii, male carapace markings

Mangrove tree crabs are difficult to see on branches in the wild. Females tend to have more and smaller spots than males.

Mangrove tree crab female holding onto a branch, side view. This species has oddly angled claws and does not use them in defense (they do not pinch fingers).

Moon Crab

Gecarcinus quadratus de Saussure, 1853

This is an intensely colorful, hardy, and massive animal that is also called a Halloween crab because the carapace can be mostly orange and black. A cage at least twelve inches (27 cm) across is needed to keep specimens alive for years, though they can be safely kept in small tubs by vendors for up to a few months. As with most species, there is no record of captive rearing. One reason has been that females are often hard to come by since they are not as brightly colored and may be passed over for the pet trade, but the main reason is tiny, planktonic larvae. Moon crabs are found in mangrove swamps, sand dunes, and rainforests across the tropical New World from Mexico to Panama. This and the following two species are members of the Family Gecarcinidae, commonly known as land crabs. They can survive with very little water as adults, but the larvae are planktonic and they must be released in the ocean to develop. Possibly the most famous member of the land crab family, Australia's Christmas Island red crab (*Gecarcoidea natalis*) is known for massive yearly migrations to the sea.

Moon crabs seem to be much stronger and live longer if given the chance to dig and build burrows. If kept in shallow water they do poorly and may begin to drop appendages before dying. If kept in deep water they die quickly. 75-100 mm (~3-4") of damp sand makes a good substrate. (Bags of play sand are readily available and inexpensive.) Any adult specimen should live at least five years in captivity with reasonable care. Mature specimens of the size commonly available in the trade molt about once every two years. It is a protracted ordeal where they dig a shelter, re-

This *Gecarcinus quadratus* (September 2015) is mostly black and orange with purple claws. Over previous decades most specimens were this approximate color, earning a different common name, Halloween crab.

Moon crab (*Gecarcinus quadratus*) adult pair together in a coconut hide. Recent imports have been a baby blue form.

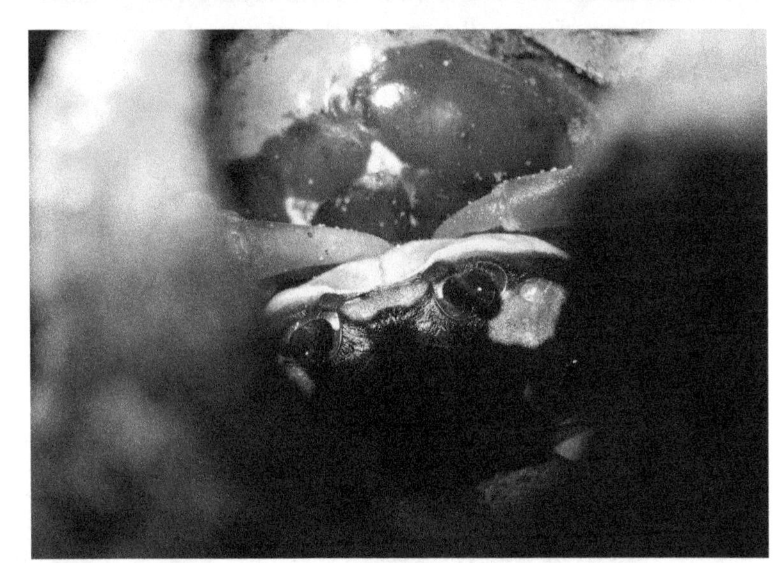

A view into the moon crab molting chamber

Moon crab next to its exuvium (molt at left). This male regenerated the right arm while the carapace was about the same size.

main stationary for two months, molt, and then take another month to harden up. For approximately four months the crab will not eat, which is one reason they do not molt more often. They will move normally if disturbed at nearly any point in this process, so a sluggish crab is dying, not molting. Full-grown crabs regenerate lost appendages, but do not seem to gain any size in molts.

Availability: This is the old-school pet land crab. Specimens have been offered at pet stores since at least the 1960s. I first saw one in the 1980s at a reptile, amphibian, and freshwater aquarium store. From the 1990s to today I encounter them seasonally at reptile shows. In 2017 I saw two at a big chain pet store.

Size: 45-53 mm across the carapace.

Feeding: This is only crab I have maintained which prefers plant-based foods over most animal matter. Pieces of coconut, apple, watermelon, carrot, etc., are eaten with gusto. Dried hardwood leaves are also eaten. Soft fruits like banana, peaches, and melon are usually avoided. Feeding usually occurs at night as they hide and try not to move when the lights are on. Unless down in a hole, they come out of hiding during the day for some animal matter. This includes strips of imitation crab (processed fish), dry dog food, and thawed, peeled shrimp.

Reproduction: The male has more lopsided and proportionately larger claws (two to three times as massive on one side for a full-grown male) and tends to be more brightly colored. The female's tail is wide and the male's tail is narrow. They mate in captivity without a recent molt. Females probably only carry eggs once a year. They must return to the ocean to release tens of thousands of planktonic larvae. Tiny crabs must return to land after a few months, because as they develop they must leave the sea or drown.

Cohabitation: Adult specimens can be kept in relatively small cages without damaging each other, though a missing leg is not unheard of. (I used to see a dozen males in a 24" x 13" (61 cm x 33 cm) temporary sales display at reptile shows in the late 1990's and I did not notice any limbs on the cage floor). Still, they cannot be kept long-term in a bare-bottom aquarium. They need deep substrate and solitude to survive the occasional molt. A freshly molted animal would probably not be killed or eaten as in many other species, but it could be buried and mangled by accident. I have maintained two males and two females in a 29-gallon terrarium (30 1/4" x 12 1/2" x 18 3/4") (77 cm x 32 cm x 46 cm) without problems during molts. The substrate is 4" (10 cm) of sand and there is an 8" (20 cm) diameter by 3/4" (2 cm) high freshwater dish on top. Coconut halves and a few pieces of bark are used to aid in burrow construction. They do not usually like to touch, but a male and female will rarely hang out together in a single coconut hide.

Patriot Crab

Cardisoma armatum Herklots, 1851

This is the Old World version of the moon crab in terms of its long-term history as a pet. It ranges across western Africa, though its most recent popular name in the United States refers to the young adult's red, white, and blue coloration. Females and larger males often lose the red and orange coloration and can look similar to *Cardisoma guanhumi*, a widespread New World species that is rarely maintained and seldom available in captivity. The patriot crab was once commonly known as the soap dish crab because specimens were packed individually in soap dishes to prevent them from hurting each other or escaping. Unlike *Gecarcinus*, this species is skittish and prefers to stay submerged underwater if given the opportunity. It can survive fully submerged for days, possibly much longer. Although it does not easily drown like the moon crab, it is not made to survive in deep water. Individuals can be kept in

Cardisoma armatum are commonly available a little bigger than a large fiddler crab because they are colorful and can be handled with minimal risk. This animal already molted twice in captivity (three molts later it is much larger and the big claw can cause serious pain).

This *C. armatum* was the unfortunate victim of a rather quick fight with a slightly larger male.

Two months later, the *Cardisoma* victim almost fully regenerated.

water just under a half inch (1 cm) deeper than their body. They do not require emergent areas when kept in shallow water. Specimens with a carapace width of an inch or greater will molt two or three times a year. The pictured specimens have molted half a dozen times over two years and have been maintained in fresh water the entire time. This species may also be kept in brackish water. From the growth rate in captivity and maximum size, specimens ought to survive at least ten years.

Availability: Patriot crabs are regularly available, but only through mail order and are usually only males. Like the moon crab, these have been imported as the rare pet since the 1960s, though I have never seen or heard of this species at a pet shop or reptile show.

The same *Cardisoma armatum* a few molts later. He usually spends the day under the coconut hide in the foreground. Before each molt the carapace become pale at the edges and brown towards the center (this is two days before a molt).

Size: Specimens usually come in less than 30 mm across, but older animals can exceed 80 mm across the carapace in nature.

Feeding: They will eat almost anything not made of metal or glass. Animal matter is preferred over vegetable, especially other crabs, crayfish, cockroaches, crickets, fish food pellets, ghost shrimp, and worms. Tiny prey items like daphnia and fruit flies are too small to be captured by available specimens. Most fruit is eaten immediately. They also consume carrots and dead hardwood leaves.

Reproduction: Females are almost as colorful as the males and may show up now and then. I have not yet been able to find a female of this species for sale. Development of planktonic larvae makes captive breeding unlikely even if a female could be acquired.

Cohabitation: Males almost immediately begin to tear each other's legs off if placed together. I made the mistake of trying this once, but fortunately the dismembered male regenerated all but one rear leg at the next molt (it did not gain any size). The molt took only two months (versus four to five months for previous molts) because molting cycles speed up when a front claw or multiple legs are missing.

Blue Land Crab

Cardisoma guanhumi Latreille, 1825

Getting one of these is like getting a cat, if the cat had one big claw, a hard shell, and should never be let loose to run around the house. Even a big, old animal should live for many years and the hobbyist should expect this to be a long-term pet. Small specimens are brown to orange, while mature specimens are lighter colored with a purple to light blue carapace. The legs can be blue, white, or orange.

Availability: Adults can be purchased from vendors on occasion, are very rarely seen at reptile shows, but never at pet shops. Though this

Blue land crab, *Cardisoma guanhumi*, male, in Costa Rica (Welmar Meneses)

Cardisoma carnifex, female, in Diego Garcia (Drew Avery)

crab can be found as far north as southern Florida, it is very rare there, especially alive. People only see them part of the year because they cross roads to breed and get smashed by cars. The rare imports usually come from Barbados or Puerto Rico where these large crabs are commonly collected for human consumption (Kick pers. comm.). The insignificant number acquired by the pet trade would otherwise end up on a local dinner plate at a significantly lower final consumer price. This species is a very popular food item in Brazil where populations may be declining (Botelho 2001).

SIZE: 110 cm (>4″)

FEEDING: Almost anything animal or vegetable is eaten. Processed fish, carrots, peaches, etc.

REPRODUCTION: Males have one big and one small claw, whereas females have two small claws.

COHABITATION: They eat each other.

Thai Micro Crab

Limnopilos naiyanetri Chuang & Ng 1991

This small freshwater crab has become a somewhat popular fare in the aquarium trade since around 2011. The lanky appendages, pale color, and fuzzy appearance make them nearly invisible when they cling to plant roots. The adult crab's body is smaller than almost any other mature crab, but the legspan is not so diminutive. A very large, old adult's carapace can be a little larger than a pencil eraser, yet the legspan well over an inch—though most never grow that big. The genus name comes from the Greek words for pool or marsh (*limno*) and wooly (*pilos*). This species has a rather impressive mat of setae on the appendages—impressive compared to many other crabs, though less so compared to some spider crabs. It is a member of the Family Hymenosomatidae of false spider crabs (currently still grouped in the spider crab Superfamily Majoidea) (Guinot 2011).

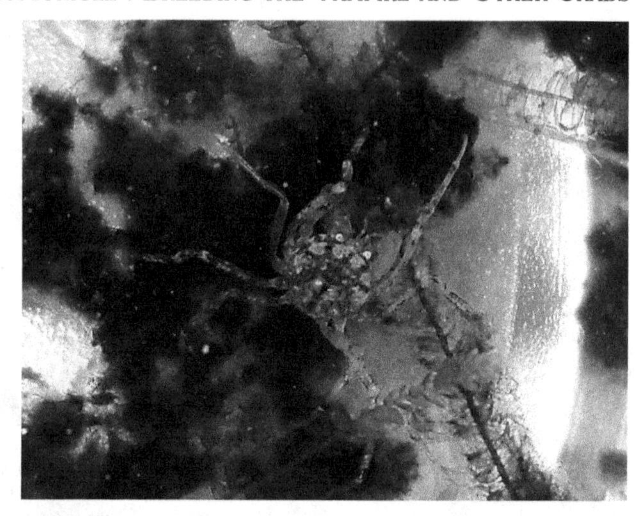

Thai micro crab (*Limnopilos naiyanetri*)

Wild specimens are found living on the roots of water hyacinth (*Eichhornia crassipes*). Unfortunately, water hyacinth is difficult to keep alive in home aquaria due to incredibly high light requirements. However, its roots resemble those of water lettuce (*Pistia stratiotes*). Water lettuce can be kept for years in aquariums with standard plant lighting (a 9- to 13-watt compact fluorescent closer than 18″ or <46 cm). The micro crabs seem to eat the *Pistria* roots, though not usually faster than healthy plants can grow. Other aquatic plants (such as Java moss) or synthetic sponges can provide a good perch and the *Limnopilos* feed readily on fish food flakes and small pellets.

This species molts at an incredible pace compared to other pet crabs. Mature specimens molt every three to six weeks (McMonigle 2014). I have kept specimens as cold as 65° F (18° C) for months, but anything below 72° F (22° C) is not suggested since they tend to get stuck in their molts when cold.

AVAILABILITY: Commonly available from multiple vendors and sold in groups of ten. Specimens are available from specialty vendors yearround, though they are not likely to be seen at any local pet store. Even if these crabs were displayed at a shop, they are too small for the casual observer to notice.

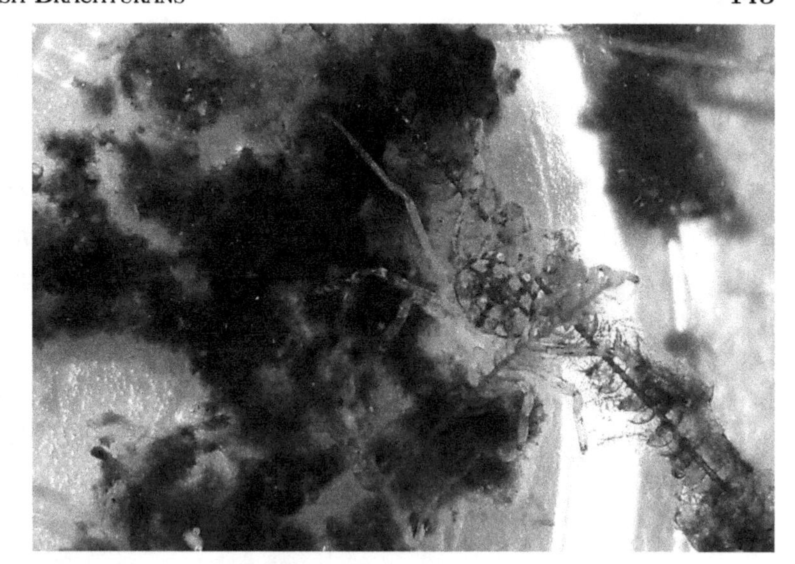

Thai micro crab (*Limnopilos naiyanetri*)

This is a very large (9 mm carapace width) adult female *Limnopilos naiyanetri*.

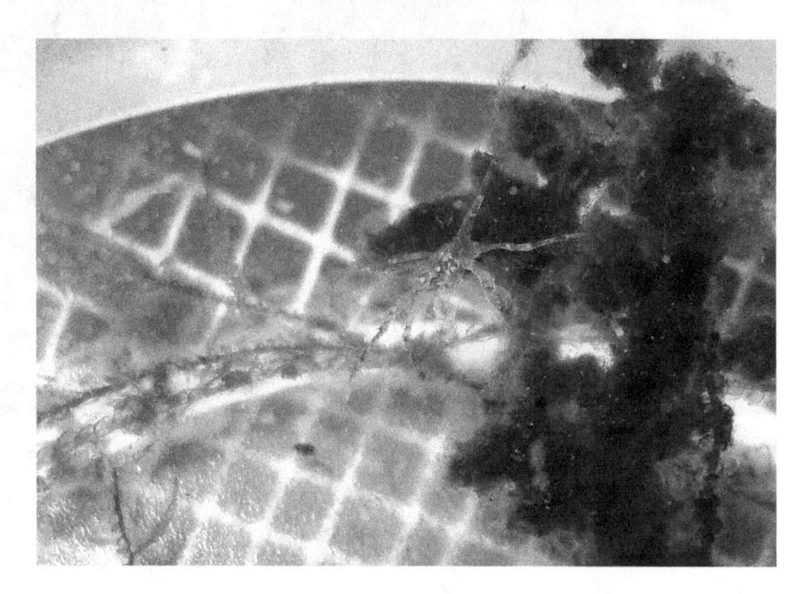

A normal size (4 mm carapace width) Thai micro crab on algae covered Java moss.

Size: Up to 8 mm carapace width, but available specimens usually have a carapace width of 4 mm or less and grow little, if at all, over successive molts.

Feeding: Certain plant roots, flake and pelleted foods. It is difficult to observe feeding due to the small size and cryptic nature.

Reproduction: Although specimens make great pets and can survive more than two years in captivity, larvae have not yet been reared in captivity. The eggs are yellow and quite notable, while the tiny babies usually swim around a few days or so before expiring from starvation.

Cohabitation: Specimens often touch and are somewhat gregarious. They molt safely in groups.

Panda Crab

Lepidothelphusa cognetti (Nobili, 1903)

This species is found on the island of Borneo in Malaysia. It is also sold as the "white arm Borneo crab" and may be labeled "Geosesarma Borneo" or sold as "*Phricotelphusa sirindhorn* (from Ngao Waterfall Forest Park in southern Thailand)." When pressed for locality and collection data, the vendor I acquired these from said "they

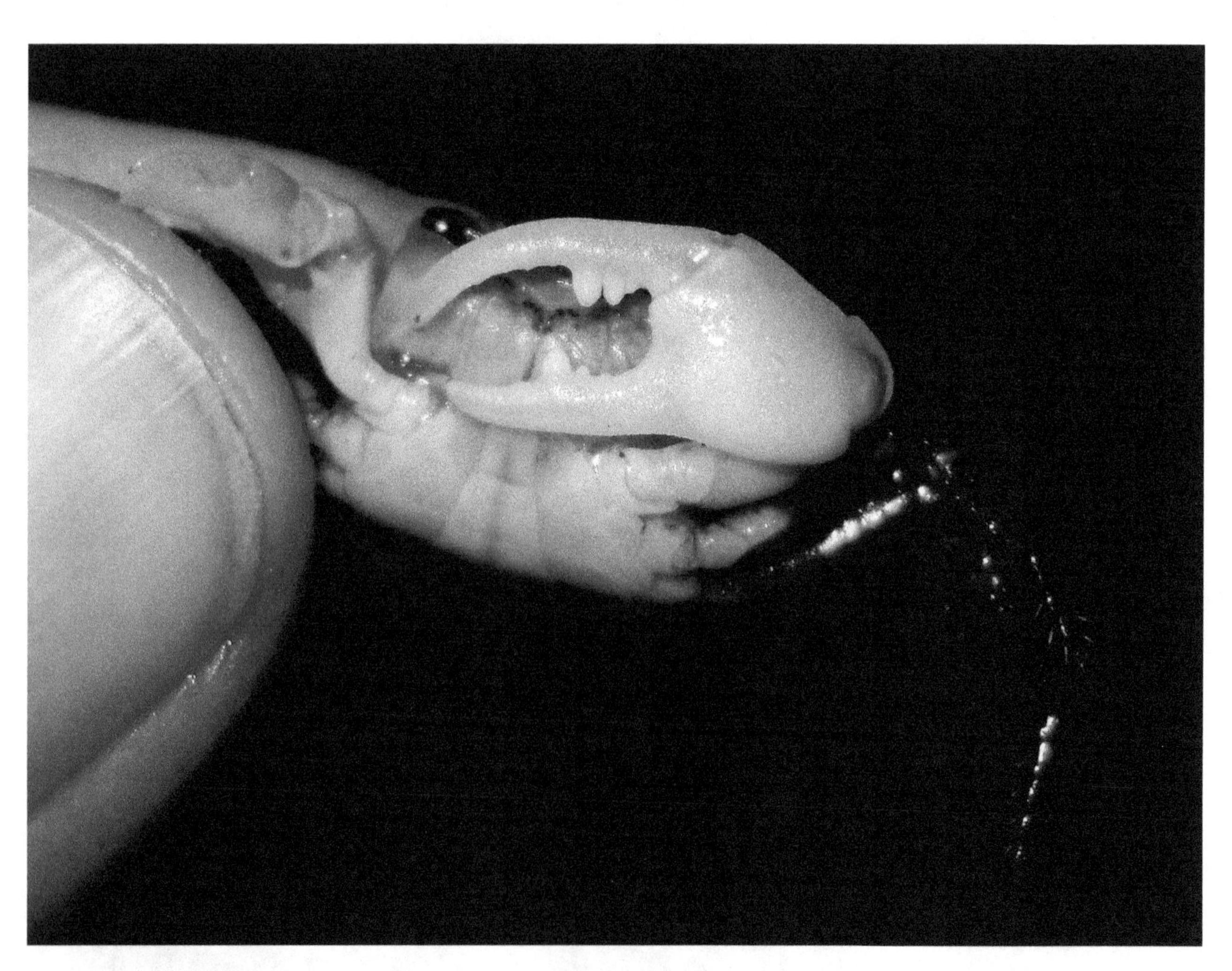

The *Lepidothelphusa cognetti* male's claw is enlarged similar to those of fiddler crabs, though it is from an unrelated group commonly known as freshwater crabs.

Panda crab (*Lepidothelphusa cognetti*) adult pair

are bred in Asia." The natural habitat is supposed to be freshwater streams, so brackish or saltwater enclosures might kill them. They prefer not to leave the water if there are no adequate hiding areas on the land portion of the habitat (the ones in the photos I moved into position above the water by hand). Of course, there is always an exception; the gravid female stays above the water or at the upper edge, very likely because the large eggs require more oxygen. Although panda crabs resemble sesarmids they are members of the freshwater crab Family Gecarcinucidae, like panther crabs. The white carapace often turns gray at night, sometimes during the day.

AVAILABILITY: Difficult to come by, even from specialty vendors.

SIZE: Males are 12 mm across the carapace (the corresponding enlarged claw is 15 mm). The mature female's carapace width is 10 mm. Specimens, especially females, look much smaller at full-size than the available species of vampire crabs.

FEEDING: They will eat small pieces of fish, fish food pellets, and daphnia, but they have a very tiny appetite. Dead leaves, fruit, and algae inspire very little feeding or are wholly ignored.

REPRODUCTION: Males are notably bigger and have one enlarged claw. Females have small front claws and a wide abdomen that covers most of the underside. Females have produced young in captivity—around 40 eggs were noted in a clutch (Höhle & Singheiser 2016). Females tend to eat little or nothing during gestation, which takes at least three months.

COHABITATION: A male and one or more females can be kept together safely.

Panther Crab

Parathelphusa pantherina (Schenkel, 1902)

This freshwater species is only known to occur in a small area of Lake Matano in Sulawesi, Indonesia. IUCN lists it as a threatened species because it is found in a very small area (an estimated extent of three square kilometers) that is under the influence of nickel mining. If the water quality becomes too degraded, the habitat will no longer support this species (Esser & Cumberlidge 2008). Collection is not listed as a concern as it can reproduce rapidly in an acceptable environment and it is not commonly eaten by humans. Captive breeding has the potential to save this creature if the natural lake habitat were to become too hostile, but I am aware of no public or private efforts to discover or establish a breeding regimen. Specimens had been coming in at approximately full-size (33-40 mm across the carapace) though lately they seem to be all immatures (around 15 mm), which may point to aquaculture farming efforts.

Like the other 1300 or so freshwater crabs, this species can be maintained in a standard aquarium because it is fully aquatic and never needs to leave the water. Freshwater crabs are characterized by large eggs with direct development (like crayfish). The problem with keeping panther crabs comes in the ease of escape. They are able to swim with some effort and can climb cords. I had an adult in a ten-gallon aquarium for

Some freshwater crabs do not require pools or streams to survive. This *Sundathelphusa* sp. finds enough water collected in tree cavities to survive. Panay, Philippines. © Christian Schwarz

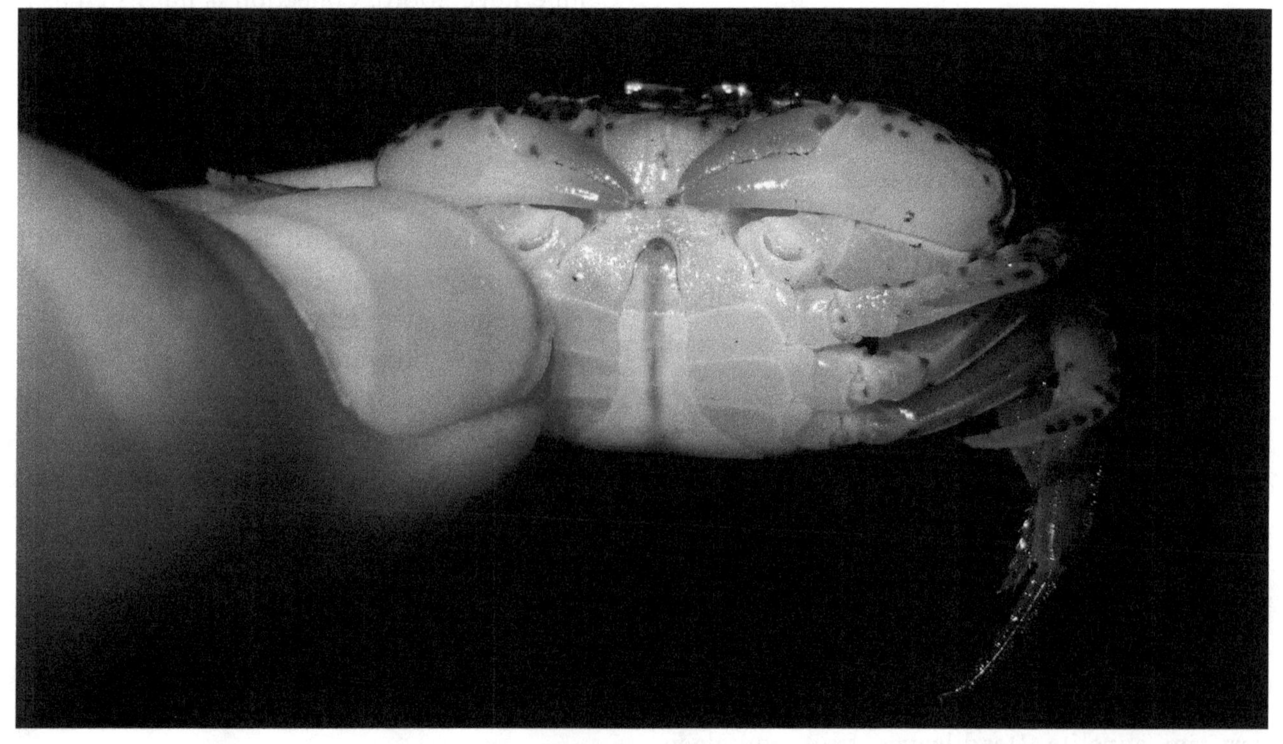

Panther crab, male abdomen (pleon)

Parathelphusa panthernia, female abdomen, a mature specimen

Parathelphusa panthernia, female abdomen

eighteen months before it chose to go walkabout. I found it the next day and it seemed strong, but it died by the next morning as the gills had dried out. They take barely six months to grow from 15 mm to 30 mm in the aquarium and should be able to live for years. Long-term success with this species in a semi-aquatic environment is unlikely without an airstone or water filtration.

AVAILABILITY: Common from multiple mail-order vendors and occasionally from local pet stores.

SIZE: 30 to 40 mm across the carapace at maturity.

FEEDING: Panthers ravenously feed on small snails, but also eat strips of processed fish and thawed shrimp when the snails are gone. Aquarium snails small enough to be crushed by the claws are killed and consumed rather quickly, but large snails may remain unharmed even if the crabs are very hungry. Thin-shelled snails like the common aquarium ramshorn (*Planorbella duryi*) are favored over thick-shelled species like the Malaysian trumpet snail (*Melanoides tuberculata*). Flake or pelleted fish food is taken with limited interest. Fish, specifically guppies, can live safely with this species in an aquarium

Panther crab (*Parathelphusa panthernia*), from June 2014 import

The left claw of *Parathelphusa panthernia* tends to be a little larger, but it is not significantly adapted for hard-shelled prey. Specimens have difficulty eating thick-shelled prey, specifically cone-shaped snails like *Melanoides* and *Tylomelania*.

indefinitely. This species also never seems to figure out that living ghost shrimp are edible. Daphnia, mosquito larvae, and brine shrimp are also not eaten as long as they are alive and swimming. I originally did not think they had much interest in vegetation because the tank had Java moss, "Asian freshwater seaweed" (*Lomariopsis lineata*), and some floating plants. However, when I introduced curly-leaf pondweed (*Potamogeton crispus*), sago pondweed (*Stuckenia pectinata*), Eurasian watermilfoil (*Myriophyllum spicatum*), and hornwort (*Ceratophyllum demersum*) collected from a nearby lake, the plants were grabbed and eaten. Eel grass (*Vallisneria americana*) was untouched. Most of the plants were coated with small zebra mussels, fly larvae tubes, and aquatic mites.

Reproduction: Sexing specimens is unusually difficult for this species unless the specimens are full-grown. I checked fourteen immatures (+ or - 1.5 cm carapace width) and they all had abdomens with a narrow shape characterized as male. I purchased three "adults" a few years back that I believed to be all male at the time. However, it seems at least one of them was a female, according to old photos. The females do not have greatly widened tails until they are sexually mature. Small specimens may still be sexed, as the female's tail is more triangular, without an abrupt narrowing near the base. Large immatures of both genders look similar, but full-grown males have larger claws. Development of the eggs is direct and there are no planktonic stages.

Cohabitation: Molting specimens can be eaten and fights may break out independent of available food and hiding spaces. Lost appendages are common when they are kept in high densities, as in a pet shop. Small specimens do not attack each other if given space and individual hiding areas, but once they exceed 20 mm across the carapace they occasionally fight. This usually results in one cheliped broken off, but not eaten, at least for a few days. I kept five specimens together safely over numerous molts for more than a year. They were kept in a large aquarium (454 liter/120 gallon with a floor space of 48" x 24" / 122 cm x 61 cm) with a number of different types of hides and still they eventually began to quarrel. The smallest was attacked and had a claw and walking leg removed. Of course, they left the guppies and ghost shrimp alone. The remaining four (2 male, 2 female) grew to 31-34 mm sexually mature specimens without incident.

Purple Matano Crab

Syntripsa matanensis (Schenkel, 1902)

This species is similar to the panther crab and is also limited to Lake Matano, but it is found over a much greater area, fifty square kilometers of the lake and surrounding areas (Esser & Cumberlidge 2008). While there are a few old images online of a yellow-legged specimen with a purple carapace purported to be this species, available specimens look very different. They are brown with brown legs and hundreds of small dark spots all over the exoskeleton. There can be purple highlights on the legs, but the purple aspect of the name is primarily a sales gimmick. *Syntripsa flavichela* from Lake Towuti and Lake Mahalona, also in Sulawesi, has made it to the aquarium trade on rare occasion and it looks similar, but has more purple coloration, larger spots, and off-white chelipeds with black fingers.

Overall care and behavior seems identical to the panther crab, but they are slower moving and slower growing. They likewise enjoy eating snails and are physically suited to consume shellfish. The left claw is about twice the size of the right and has large white molars to crush hard-shelled prey. In a year they molt once or twice, when a panther would molt three or four times.

Availability: Matano crabs became commonly available from mail order vendors in North America starting in 2016. This species is eaten by humans and might have seen the rare delicacy import, but the first live animals for aquaria were imported into Germany in 2007 (Fritzsche 2008).

Size: Carapace width might reach four or five centimeters, but the largest I have on hand is 24 mm, while another grew one millimeter after a molt (from 22 mm to 23 mm). Available specimens seem like they would never be large enough to be worth eating.

Feeding: Matano crabs feed less voraciously than panthers and likewise prefer small, thin-shelled snails to other foods (the snails must be small enough to fit in the crux of the claw). They will eat strips of processed fish and pelleted fish food if very hungry. They are not aggressive feeders.

Like *P. pantherina*, the female does not have a wide tail even when she is relatively large, though this is probably different for sexually mature crabs. Males and females have similar-sized claws and are approximately the same size. Development of the eggs is direct and there are no planktonic stages.

Purple Matano crab (*Syntripsa matannensis*) female abdomen (pleon).

Purple Matano crab male abdomen

This Purple Matano crab is missing a rear leg. Brown overall color, peppered with small black spots, is the normal coloration for imported specimens of this species.

Syntripsa males and females have a large smashing claw on the left side with large white molars and a smaller, cutting claw on the right.

This is the same female as opposite bottom left, after molting and regenerating the back right leg. The coloration seems to have darkened slightly.

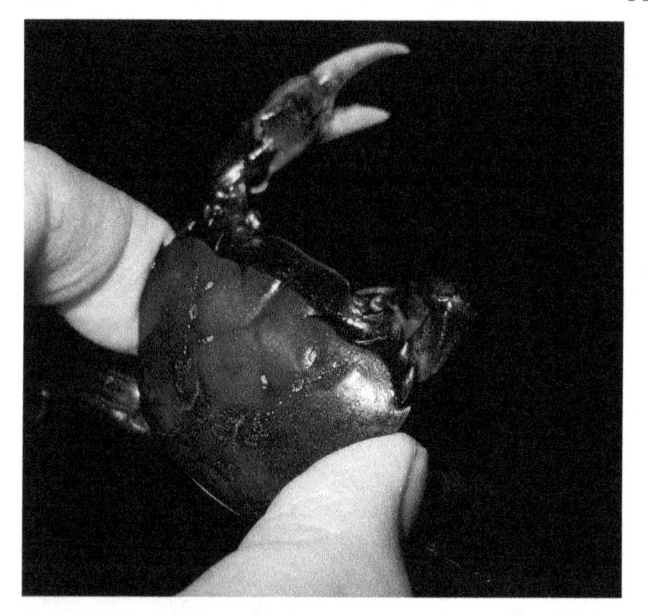

Somanniathelphusa sp., from Vietnam, carapace

Somanniathelphusa sp. from Vietnam really can be this purple, though some are brownish.

Cohabitation: With adequate space, they do not bother each other, except possibly during molts.

Vietnamese Purple Barking Crab

Somanniathelphusa sp.

This genus is from the Family Gecarcinucidae (the same family as *Lepidothelphusa*, *Parathelphusa*, and *Syntripsa*). The species pictured is among dozens of tropical Asian species known as rice paddy crabs because they are commonly encountered in paddy fields. Paddy crabs are very common food items in Asia and many specialty meals require them as an ingredient. They can probably give a good pinch, but they are easy to grab by the sides and hold without getting pinched.

The species pictured can be entirely brown, but specimens are often a spectacular purple color (two of five females acquired, three of which were alive). After observing a number of molts, I learned the color can have as much to do with molt proximity as individual variation. One brown specimen turned purple after the molt and retained a good bit of purple color for many months following. Another molted and remained brown on the carapace, but the ends of the legs turned purple. The purple ones are much deeper purple after a molt, but fade somewhat and can become mostly brown just before the next molt. Large specimens can molt twice in as little as three months.

Adult females produce impressive barking noises through stridulation. Noises come from within the carapace (possibly from below the eyes). It is not possible to see appendage movement associated with the noise. Stridulation can be stopped, at least temporarily, by bumping the aquarium or touching the crab. The noise can be heard from around 20 ft. (6 meters) away and sounds like a creaky door. It is rarely heard but when it occurs it continues on and off every few seconds for up to a few hours.

Availability: Almost never seen on North American vendor lists.

Size: Paddy crabs are large animals compared to any other available freshwater (or

Somanniathelphusa sp. can live and grow in deep enclosures with no access to land. This was originally a small brown specimen, now on its second molt. It acquired some purple color, but very little compared to the bright purple specimen that was next to it.

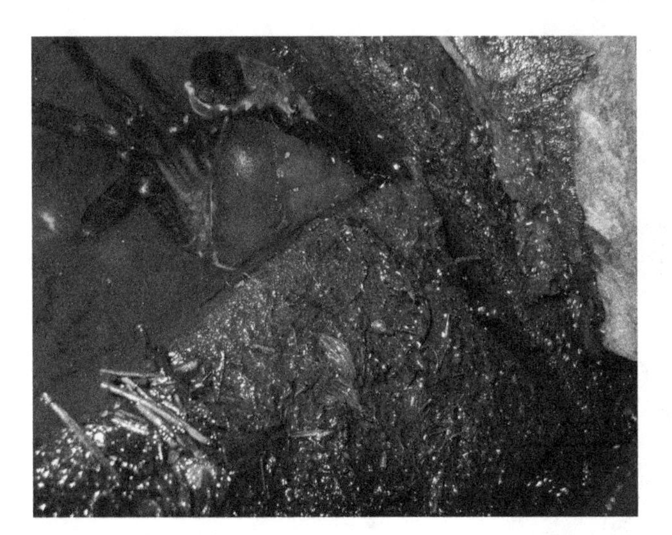

Somanniathelphusa may fare better in shallow habitats with emergent areas.

brackish) species and can grow to at least 54 mm across the carapace.

FEEDING: These are good feeders that readily consume processed fish, frozen shrimp, snails, earthworms, and fish food pellets almost daily. Tadpoles and fish are caught in shallow water enclosures, but may avoid predation for long periods in deep water. Ghost shrimp avoid predation until the crab realizes they are food (which can take weeks or months), then the crab eats every last one in a day or two. Various aquatic plants, carrots, hardwood leaves, apples, grapes, strawberries, and peaches are eaten. They also chew up hair algae, Java moss, and artificial or real sponge, but gain little or no nutrition from these items.

REPRODUCTION: Females have a widened tail to gestate the large eggs. Development of the eggs is direct, so they enter the world looking like a small adult crab.

COHABITATION: Probably like *Syntripsa,* but I have not tried keeping them together. They cannot be kept with any plants or most animals, but they do not bother fish that live in the upper water column (specifically guppies). I kept one in a 10-gallon aquarium with two bamboo shrimp for six months without predation.

Fiddler Crabs

Uca spp.

The namesake of the fiddler or signal crab is the mature male's single, gigantic claw. This oversized monstrosity is displayed to other males

This *Uca crassipes* was placed in a reef aquarium and lived less than a month without access to land.

Uca minax male fight damage includes a missing cheliped.

Uca minax major male.

Uca pugilator during a mating attempt.

Uca minax with eggs appear grayish when ready to hatch.

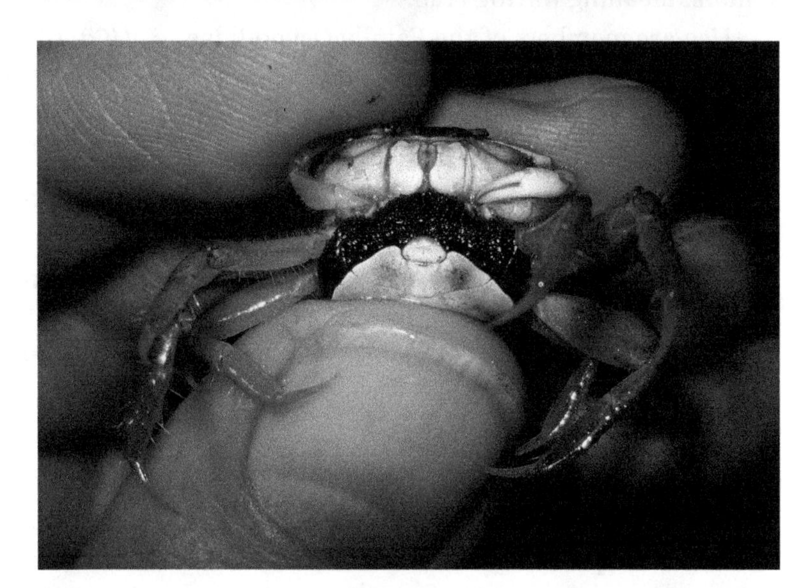

Uca minax freshly laid eggs are dark, almost black, with a reddish tinge.

Uca minax hatchlings do not look like much even with a 10x loupe.

in fights over territory and waved to attract females. The gaudy claw is usually bright white or light yellow and is used like a signal flag. However, it is not visible from every direction. The enlarged claw is often dark on the upper edge to make it invisible to birds flying overhead. Males often wave the claw up and down when the keeper approaches. The English name is fiddler crab (French = *crabe-violoniste*; Spanish = *cangrejo violinista*) because it looks like the male is carrying around a big violin and playing it in silence when the small cheliped moves up and down during feeding. The German name is *Winkerkrabbe*, meaning waving crab.

Uca are members of the Family Ocypodidae, which includes the ghost crabs. Not all species possess the enlarged male claw. Also, the rare female can possess an enlarged claw. They are primarily found in shallow brackish water but some species survive in full salt or fresh for long periods. They can survive submerged in water for weeks, rarely months, but eventually drown without access to air.

Red-Jointed Fiddler Crab
Uca minax (Leconte 1855)

Males of *U. minax* have orange to red joints on the large cheliped. They do not have patterns on the carapace or the purple markings usually seen on the carapace of *U. pugnax* or *U. pugilator*. This species has long been a common sight at pet stores, while it naturally ranges the coastal areas from Massachusetts to Texas.

I fell in love with these little crabs at the local pet shop when I had to be lifted by my father to see into the aquariums. When I was eight, I laminated a beloved fold-out poster from a National Geographic magazine that had painted images of mud flats and fiddler burrows to hang on my wall. I recreated the image in a 10-gallon (38 liter) aquarium with high-content yellow clay soil and stone. Brackish water

Uca minax female in a freshwater setup. This specimen lived for three years and molted several times.

was added (1.010 specific gravity using artificial seawater mix and tap water). The clay did not become mud (it was primarily impervious to water), but the setup would result in very milky water after any disturbance. Every three or four months it was necessary to dig up the entire artificial mud flat to make sure anything was still alive since the fiddlers always stayed hidden in their tunnels. The crabs would barely live a year and I do not remember any molts. Also, evaporation and water changes made it difficult to keep the salinity in check. Fast-forward a few decades and my methods have changed entirely. I now use sponge, Java moss, and dechlorinated tap water (pH 7.4, hardness 120 mg/L, alkalinity 90 mg/L). The sponge provides emergent areas and places to hide. The new setup does eliminate the fantastic tunnels created by the crabs (they had been like a massive ant farm, except there was almost no visibility) but the crabs come out of hiding regularly, molt every three to five months, and live two to three years. I was unable to get females to produce eggs in previous brackish water setups but have seen mating and eggs in recent setups. Females are moved to shallow saltwater after the eggs start to develop.

Availability: This is the most common pet true crab in North America (probably not more common than the purple pincer hermit). Fiddler crabs have been seen at local pet shops since the 1960s but artificial seawater was not readily available till the late 1970s. Butler (1856) mentions fiddler crabs in an early aquarium book, but does not discuss their captive care in aquaria: "He is astonishingly nimble, though he does make such odd progress, and he darts down his hole in the sand with a celerity that is marvellous."

Size: Specimens are usually from 15-22 mm across the carapace. The 22 mm male pictured has an enlarged claw that measures 37 mm, including only the chela of the cheliped.

Feeding: They eat pieces of fish, fish flakes and pellets, earthworms, and mud, but seem to have little interest in much else. They can produce a lot of frass from eating substrate, but trample it down and eat it again later. The temptation to feed as much as specimens are willing to eat should be avoided as this will lead to rancid conditions that can kill them. A few pellets or a tiny piece of processed fish once a day is more than enough.

Reproduction: If the oversized claw was lost in shipping or a fight, the male is easy to identify by the narrow abdomen. Adult females mate without a recent molt. I have observed mating during the day in shallow water that does not cover the crabs. Approximately ten days after mating, females brood massive clusters of tiny eggs between the abdomen and carapace. Females usually only produce eggs once a year in captivity. The eggs start out reddish black and transform to light gray, with a greenish or bluish tinge, before hatching. At 74° F (23° C) eggs of this species hatch after seventeen days. Tens of thousands of eggs hatch into tiny black spots

Mating *Uca minax*: the male is protecting his female from the camera with his large claw.

Uca minax major male. This specimen lived and grew in freshwater for two years as an adult. He was then placed in brackish water and died after four months (probably due to old age rather than the change in salinity). Quality emergent areas are key to keeping fiddler crabs alive.

Uca minax males can pinch but are usually safe to handle (unless you grab onto the claw, which quickly autotomizes).

Uca crassipes females need an emergent area for long-term (a few years) survival.

Uca pugilator females sometimes have a purple spot on the carapace, though it is more common on males during breeding season.

Uca pugilator, major male

smaller than baby brine (*Artemia salina* nauplii). They do not have the same jumpy movements, but move like seed shrimp. The planktonic crabs can be offered cultured rotifers after a few molts, but usually die the third day after hatching.

Werner (2003) says (referring to *Uca* spp.), "Hence rearing them would be virtually impossible." Captive reproduction is certainly easier said than done, but it is possible (Rupp 2004). Halton (2013) says a number of internet sources claim to have bred fiddlers in home aquaria, but never with developmental photographs or other photographic documentation. This is true for standard home aquaria, but breeding in a home aquarium and providing a plankton kreisel tank with cultured rotifers, green water, and baby brine shrimp has been documented photographically for some *Uca* species, while unsuccessful for others (Rupp 2013).

Cohabitation: They are usually kept in groups but mature males can fight, damage, and kill each other (the winner does not eat the loser, he just maims him). Over many months, all but one male in the enclosure often ends up being killed, though providing a large cage (bottom dimensions greater than 30 1/4" x 12 1/2" or 77 cm x 32 cm) reduces aggression. If five or more males are kept in a large enclosure, they set up spaced "waving stations" and aggression can be spread.

Uca pugilator with freshly laid, burgundy-colored eggs

Uca pugilator Bosc 1802 is known as the Atlantic Sand Fiddler and is found from Massachusetts to Florida (Kaplan 1988), the Florida Keys, and the Bahamas. It is much prettier than *U. minax*, with contrasting coloration and geometric markings on the carapace that tends towards pale white. Males, less often females, in summer have purple markings on the carapace. I have never seen this species in a pet shop, but they can be ordered as fishing bait. Unfortunately, bait specimens have an initial die-off that can be greater than 50% in the first week and 75% after two months. In captivity, they are very similar to *U. minax* in behavior and feeding. Brackish water can be used to keep specimens alive up to a few years; fresh water has not been tested over long periods. The largest male I have measured had a carapace width of 20.5 mm and a front claw (last two segments of the front leg only) length of 38 mm.

Reproduction: Males make a loud drumming noise (three to five rapid taps in succession) every so often, usually during daylight hours. The sound can be heard from across the room and is probably employed to attract females. The large claw is rapped against a surface such as a piece of wood or stone, or the side of the enclosure. I once observed a male drum against a piece of sponge and of course no sound was produced, but it seemed like he was testing different surfaces to test for the best sound production. Mating is not commonly observed, though it occurs during daylight in shallow water. Eggs take nearly double the time *U. minax* eggs require to develop. They change from reddish black to greenish gray and hatch after about thirty days (McMonigle 2017b). Though incredibly tiny, only about a fifth of a millimeter across, they are much larger than hatchling *U. minax*. I had success rearing them

Watermelon crabs, *Uca* sp. in salt water with a Malaysian gametophyte fern that resembles macroalgae. The male towards the front lost his enlarged claw in shipping.

Sometimes the wavy blue lines are green so the body of this crab resembles the pattern of an incredibly tiny watermelon.

Uca sp., Thailand (Rushen)

Blue fiddler crab, *Uca tetragonon*, Malaysia (Bernard Dupont)

Calling fiddler crab, *Uca vocans*, Malaysia (Bernard Dupont)

Ring-legged fiddler crab, *Uca annulipes*, Malaysia (Bernard Dupont)

Red fiddler crabs in the Sunderbans (Sayamindu Dasgupta)

Yellow fiddler crab, India (Lalithamba)

Uca sp., Hong Kong (Sarah Joy)

Uca sp., strawberry fiddler

Uca splendens male

through the first four instars with live green water and rotifers. A few specimens were observed feeding on decapsulated *Artemia* eggs from the second zoeal instar forward.

Uca crassipes (White, 1847) is a tropical Asian species found on the shores and islands of Australia, Japan, New Guinea, Philippines, Taiwan, and Thailand (Shi et al. 2016). It is commonly seen at marine pet shops and sold as "reef safe" since it does not bother corals, fish, or other large invertebrates. Unlike the North American *Uca* listed above, which can eat large quantities of select food, this species eats almost nothing and yet it can live just as long. A tiny piece of shrimp may be eaten every two weeks; if fed more often, food will be ignored and rot. Adult brine shrimp are also eaten. Specimens rarely survive a few months fully submerged in a reef tank, but they will live at least a year when given emergent areas and brackish water. They are extremely shy, often only climbing onto emergent areas in the middle of the night. When handled, they can pull in their legs and remain motionless for up to a few minutes. Female specimens are usually sold as "red reef crabs," not fiddler crabs. I have seen hundreds in pet shops and have never seen a male, possibly an attempt to keep them from being called fiddler crabs.

Uca splendida (Stimpson, 1858) is found around China, Taiwan, and Vietnam (Shi et al. 2016). Males (only) of what appear to be this species have shown up from specialty vendors in North America on occasion, labeled as freshwater Watermelon Crabs. However, the source for this material is supposedly Malaysia. These are hardy and easy to care for, but expensive and difficult to locate, even from specialty vendors. They can live a year in brackish water as long as there is access to air. Fresh water may or may not result in long-term survival. Like *U. crassipes* they desire and require very little food, but feed well on adult brine shrimp. Size: (from two wild males) carapace width 18-19 mm, claw 28.5-33 mm.

The Strawberry Fiddler is a recent import of male-only specimens from tropical Asia. It does not have white spots like a strawberry hermit—rather, specimens are pink with white-tipped appendages. Other imports from South America pop up on occasion and resemble *Uca pugilator,* but with a different pattern of markings on the carapace.

Ghost Crab

Ocypode quadrata (Fabricius, 1787)

In 2010 I visited a local beach in Norfolk, VA. There were a number of people fishing, and a few joggers, but it was very different from the sterilized, synthetic, public beach in nearby Virginia Beach. Bunches of washed-up seagrass dotted the sandy shoreline, along with the occasional large clumps of dead *Gracilaria* macroalgae. Sea lettuce (*Ulva lactuca*) was mixed in with sea grass. The beach was mostly sand, but larger gravel accumulated where the water hit the shore. It looked like "natural" freshwater aquarium gravel. I found two small stones with an interesting green macroalgae attached, *Enteromorpha flexuosa.* (After returning home I remembered why I never keep *Ulva* or *Enteromorpha.* The blue-legged hermits employed to keep down hair algae make quick work of these plants. They ignore most types of macroalgae, including *Caulerpa* and *Valonia.*)

About fifteen feet from the shore I notice holes from 0.5″ to 1.5″ (13-38 mm) in diameter. I had seen similar holes before on beaches, and digging up the area, had never found a thing. That did not stop me from digging. I offered my daughters and oldest niece $2 to figure out what was in these holes. For all I knew, the owner was

Ocypode habitat, Norfolk Beach, Virginia

Ocypode burrow in nature

Ocypode burrows in captivity cannot be deep. Water does not flow through the sand like it does on a beach, so depth would result in anaerobic conditions.

a mouse, snake, or fairy. We all worked on our own area for ten minutes when I noticed a little white figure shoot out of the shallow pit my nine-year-old was digging. I watched it head straight for the water. Two seconds and fifteen feet later I grabbed it from the crashing waves. It gave me a pinch, but it did not hurt much nor draw blood. It was a handsome, light-colored land crab with a sandy, dappled carapace and very tall eyes. It was about twice the size of a fiddler crab and appeared to be a male from the narrow tail folded against the body. Many holes later I found a second crab, a little bigger. We put them in one of my 3.5-year-old twin nieces' play buckets with a little water, and the youngsters temporarily took possession of the crabs. A third male crab pinched me and left its claw hanging in my skin with a little blood, so I let it go to regenerate its arm.

Ocypode quadrata is easy to identify since it is the only ghost crab found along the Atlantic coast of the United States. A few dozen other species are found in the genus worldwide and are also called ghost crabs. Specimens have one large and one small claw, but these are not highly exaggerated like fiddlers. They can be left handed or right handed. The carapace of the smaller specimens was dappled and looked like sand, whereas the largest four (including the lone female) had a solid grayish carapace. This species feeds on washed up plants and invertebrates and eats a variety of things in captivity, from dog food to live crickets.

In the morning, with the promise of $2 a crab, my daughters came back with me to the beach. I wanted to find a female. It did not take long to find the first crab. The holes went off in all different directions (I thought they were going towards the beach until I tested that theory), but never straight down. The tunnels were only about eighteen inches deep, but the problem is

Ghost crab (*Ocypode quadrata*), average-size specimen dug out of a beach. A few specimens were the size of moon crabs.

once you lose track of the hole you lose the crab. Dig all you want and you will not find it. A shovel may have made digging easier, but most likely would have ended up with the crabs chopped in half or buried in a pile it would not dig out of till after we left.

We ended up finding eleven more crabs of which the second to last was a female and the last a monster male. He was the size of a mature moon crab, but with thicker, larger legs. His hole was close to twenty feet from the water but he made it to the surf. These crabs are crazy fast and I knew this one might do more than draw a drop of blood. When ghost crabs hit the water they only go about two feet in and stop; if they kept going we would have seen half as many crabs. The pinching claws are difficult to avoid in the water. I pinned him with a toy shovel and picked him up. When I was about to put him in our bucket of crabs I notice a few separated legs. A large male with a 38 mm carapace was angry. The smaller ones showed little aggression, but I should have anticipated this.

Overall there was one female out of fourteen crabs. Her claws and build were similar to the males, but her tail was twice as wide. We tried around sixty holes and only got one out of four overall, but three of the last five as practice made it easier (McMonigle 2010). The beach was a few miles long and if I could have stayed for a few weeks I would have dug up a few thousand holes to see if the high male to female ratio was consistent.

Ghost crabs are beautiful, common eastern North American crabs, but a feasible method for keeping them long term may not be possible. Unfortunately, this crab genus should not be kept in captivity because while they probably live three years in nature, they are difficult to keep alive in captivity for more than three months. Freshwater is bad for them and they cannot survive submerged in saltwater for more than a day or two. With damp sand and saltwater they live a lot longer, but certainly not their normal span. They also are heavy eaters, produce tons of frass, and poison their cage easily. Other invertebrates found in sandy beach habitats, including burrowing clams and mole crabs, are also difficult to keep alive.

Availability: *Ocypode* can be easy to collect from some beaches, but are not available from vendors because they die too quickly. They do not do well in captivity.

Size: Up to 50 mm across the carapace.

Feeding: Fish flakes and pellets, shrimp, crickets, cockroaches, dog food, leaves, fruit, etc.

Reproduction: Females and males look similar except for the abdomen. Planktonic larvae.

Cohabitation: Ghost crabs do not share burrows except for mating.

Dwarf Mud Crab

Rhithropanopeus harrisii (Gould 1841)

This small species is native to the Atlantic coast of the Americas from Canada to Mexico, though humans have introduced it nearly worldwide. Members of the Family Panopeidae (and other families formerly included in the Xanthidae) are known as mud crabs because they inhabit soft-bottom habitats. This species is usually

Dwarf mud crabs can become very light in color to match the background after one or two molts.

A mature dwarf mud crab (*Rhithropanopeus harrisii*)

marine or brackish, but it can be maintained and successfully grown from a tiny crab to maturity in fresh water. I have grown up a number of specimens in shallow-water caging with sponge islands.

AVAILABILITY: This is the only species from the Family Panopeidae with somewhat regular availability in the North American pet trade (commonly available from a single vendor).

SIZE: Available specimens are usually 4-10 mm across the carapace. Females can be gravid at least as small as 9.5 mm. The largest adult I have measured is 12 mm, but this species may reach 20 mm.

FEEDING: Pellets, fish flakes, processed fish, daphnia, adult brine shrimp, and mosquito larvae are preferred foods. Leaves, fruits, and vegetables may be eaten if they are very hungry.

REPRODUCTION: Adults mate during the summer months in nature; females do not molt before mating. Genders look similar and can be hard to tell apart. The adult female's abdomen is significantly wider than the male's, but it is triangular and reduced in size. Up to four broods containing 1200-5000 eggs are produced each summer (Turoboyski 1973). Larvae are planktonic. It is usually a brackish crab but it is more easily maintained in fresh water. This species has been shown to reproduce in freshwater reservoirs in Texas. The large reservoirs have enough of a food web to support the planktonic larvae (Boyle, et al. 2010). The first instar hatchling is difficult to see with the naked eye, but its body is the mass of a third instar *Uca pugilator*. The first instar zoea has fantastically long lateral, dorsal, rostral, and subocular spines.

COHABITATION: Specimens often hang out within arm's-length and usually do not harm each other. Freshly molted animals are sometimes harmed, but not eaten by their cagemates.

Black-Fingered Mud Crab

Panopeus herbstii H. Milne-Edwards, 1834

This is the largest North American mud crab and one of the most common. It may be encountered in muddy estuaries, usually around oysters and mussels. It is easy to collect when the tide runs out by flipping over stones. *Panopeus herbstii* is found along the coast from Massachusetts to Brazil (Gosner 1978). Immature crabs (15-20 mm) reach full size in two years at room temperature and full salinity (~1.022 specific gravity). Once they obtain full size, they only live another eighteen months. These crabs can be kept in brackish water (1.005 to 1.010 salinity), but probably would not survive long-term in fresh water. I kept this species in a standard marine aquarium with an underground filter, lift tubes, powerheads, and external protein skimmer. However, in the first few months some escaped and dried up in various first-floor heating ducts.

Panopeus herbstii specimen collected from beneath rocks in a mudflat in Chesapeake Bay (Virginia) July, 2009. This photo was taken in Dec. 2011.

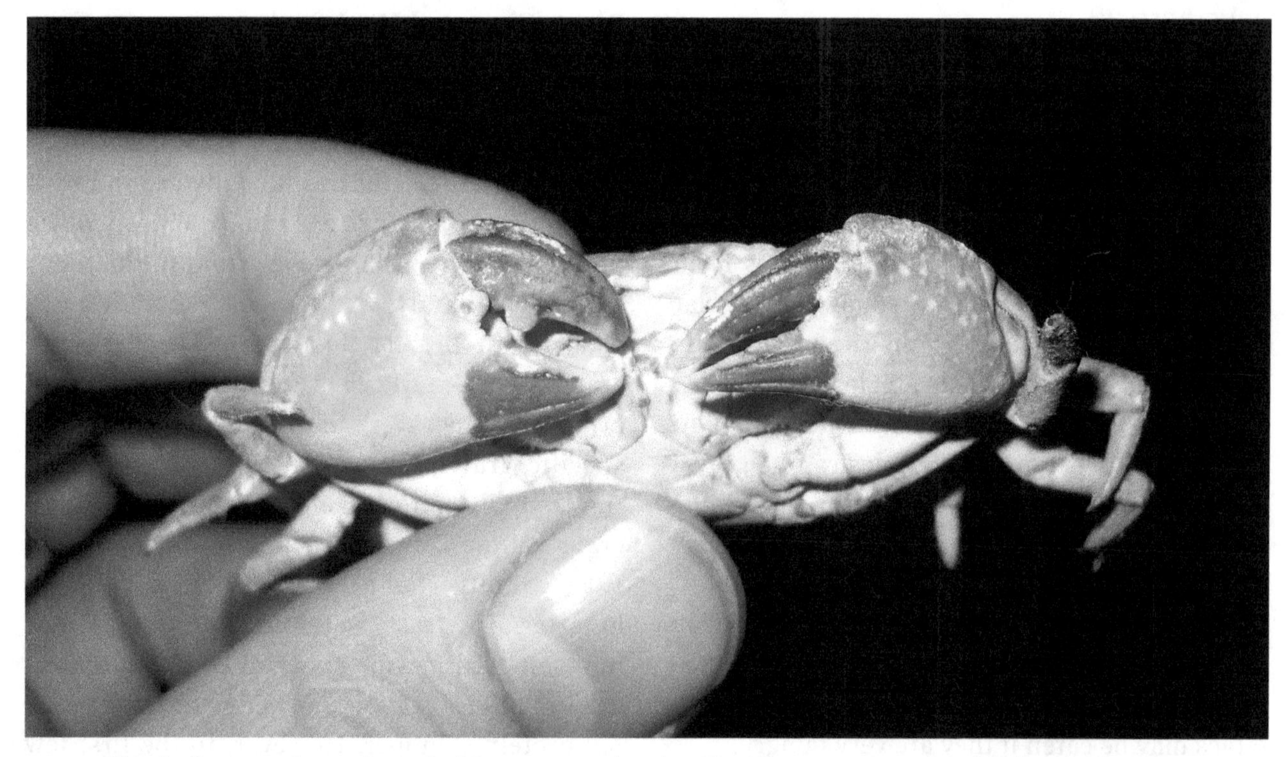

Black-fingered mud crab (*Panopeus herbstii*), the exuvium is thick and heavy.

Availability: Easily collected, but not currently offered by any vendors.

Size: Full-grown specimens are approximately 40 mm across the carapace.

Feeding: Fish food pellets, krill, shrimp, processed fish, mussels, and snails including turbo snails. Large specimens can eat two or three mussels a week (mussels marinated in a jar can last months in the refrigerator after opening). It cannot catch healthy fish and, like nearly all crabs, it does not feed on anemones or corals.

Reproduction: Males and females look similar except for the width of the abdomen. Mating takes place without a recent molt. The thousands of eggs held by the female are similar in size to brine shrimp eggs.

Cohabitation: I have seen up to half a dozen under one rock in nature, but they do not hang out in the same burrow. These crabs do not seem to harm each other in captivity, but they attempt to escape when there are multiple specimens in one aquarium.

MARINE AQUARIUM CRABS

There are countless fantastic crabs seen at marine pet stores for the home aquarium. These include brightly colored fiddler crabs (*Uca* spp.), Hawaiian strawberry crabs (*Neoliomera pubescens*), calico box crabs (*Hepatus epheliticus*), and pom-pom crabs (*Lybia* spp.). *Lybia* are popular because they hold tiny anemones in their front claws. Some marine crabs are kept in reef tanks or fish-only aquariums, but it is important to remember copper-based medications for fish diseases can kill crabs and other invertebrates within minutes. The following selected crabs are extremely common in captivity and have been for decades. The last two require marine aquariums, but are not seen in aquarium shops.

Sally Lightfoot Crab

Percnon gibbesi (H. Milne-Edwards, 1853)

This is an old aquarium species that is widespread across all tropical New World coasts. It was recently introduced to the Mediterranean Sea. Divers call this species the urchin crab because it is associated with long-spined sea urchins. The Sally Lightfoot crab is easy to keep in marine aquaria. Although it is found near shore, it is not a semi-aquatic species and it is very unlikely to escape a normal aquarium. They molt regularly in captivity, once every two to three months, and live at least three years.

Availability: Regularly seen at most marine pet shops.

Size: Carapace width reaches 20-24 mm and is usually 10% longer than wide.

Feeding: They have a good appetite for pelleted or flake fish food and pieces of fish, shrimp, or mussel, but plant materials are generally ignored. Anything that moves is also ignored. My species tank for *Percnon* was eventually overrun with thousands of scuds since these crabs do not eat them.

Reproduction: Males and females look similar, but the male's claws are about twice the mass of the claws on a similar-sized female. As with most true crabs, the female has a wide abdomen used to hold the eggs while the male's abdomen is narrow. Pairs often hang out together and the male may sit over the female. He moves his claws up and down in courtship dances with the female. Females can produce eggs every other month at least three times in a row. The eggs are perfectly hidden beneath the female's abdomen with no gaps or bulging to give away their presence. It would be impossible to tell when a female is carrying without tearing open the abdomen, but the bright-orange egg mass is partly visible through the underside. As with many marine species, the planktonic larvae have been wild-collected and kept through stages in the laboratory, but may have never been reared directly from eggs to the crab form.

Even hidden eggs may be visible through the abdomen, reddish egg coloration shows through on a gravid female *Percnon*.

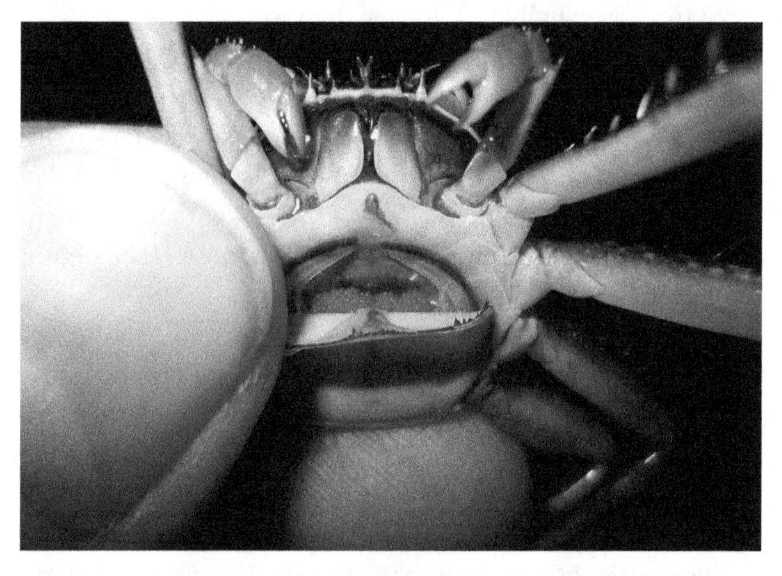

The abdomen can be pulled down to view eggs, but forcing the tail open can damage the crab.

Percnon gibbesi: the male in front has larger claws.

Sally lightfoot crabs do not burrow but they need a substrate to hold onto.

Percnon sp. (Silke Baron)

Cohabitation: Specimens of both genders and different sizes cohabitate and even molt together safely.

Yellowline Arrow Crab

Stenorhynchus seticornis (Herbst, 1788)

This common aquarium pet eats fish food, lives a few years without special care, looks spectacular, and is relatively harmless to cage decorations and fish. While an enjoyable pet, the ones available in stores are usually adults and rarely live more than a year. They tend to become covered in algae and rarely molt in an aquarium. The legs break off if there are large, aggressive fish in the tank. Even large clownfish can damage them.

Availability: Commonly available at shops that carry marine livestock. This is one of the classic pet crabs. Arrow crabs have been available to marine hobbyists at local pet stores since at least the 1970s.

Size: One of the largest females I have had was 17 mm across the carapace and 51 mm from the abdomen (folded) to the tip of the rostrum.

Feeding: Specimens should be hand-fed a piece of shrimp or fish at least once a week for long-term survival. These crabs are very unlikely to grow or survive more than a year on fish flakes alone unless no other creatures are in the setup. The legs have taste sensors which are easily evidenced when flake food is offered. As soon as a piece of fish food hits a spot on the leg, the whole crab reacts quickly to the direction and location to catch the food. The rostrum (horn on the head) has no equivalent sensors since the animal does not react when the rostrum touches fish flakes. Arrow crabs are popularly offered as a means to control bristle worms (spiny venomous polychete worms that infest marine aquariums from time to time). The arrow crab is known for eating fish on occasion, but marine fish die a lot and there is likely some cause-and-effect confusion. Nevertheless, at least one writer claims to have observed a crab using its spiny rostrum to spear a live fish (Debelius and Baensch 1994). In the ocean this species is often associated with sea anemones (Order Hexacorallia), while

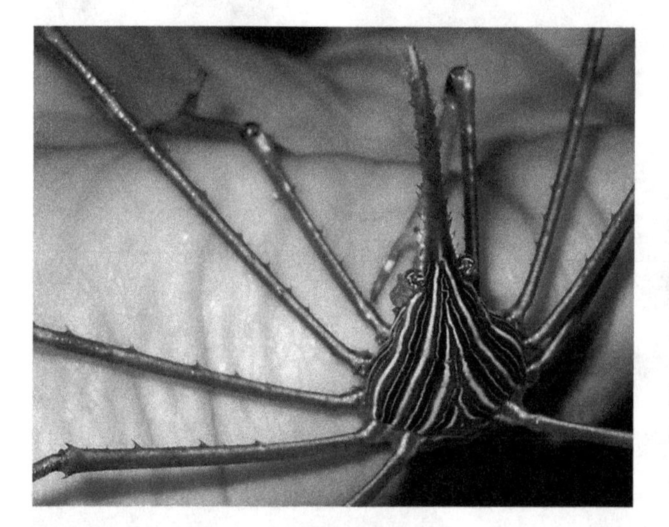

Arrow crabs are very colorful, though the carapace tends to get covered with algae over time in a home aquarium. This specimen is a recent import.

Arrow crabs are not harmed by anemones directly, but they should not be kept in a cage covered in anemones. The anemones catch most of the food and the crab may eventually starve.

Yellowline arrow crab (*Stenorhynchus seticornis*), large adult

other species may be found near urchins or tube anemones (Order Ceriantharia). Arrow crabs do not bother anemones, corals, urchins, or most starfish. However, they may bite off serpent star legs and are surprisingly adept predators of small hermit crabs. Arrow crabs will eventually eat all the cleaner snails (turbos, etc.) in the aquarium.

REPRODUCTION: If kept as pairs, the female can release planktonic larvae as often as once a month.

COHABITATION: A male and female will cohabitate nicely, but specimens of the same gender fight unless the aquarium is very large.

Decorator Crab

Camposcia retusa (Latreille, 1829)

This spider crab from Indonesia has been regularly available since at least the 1980s. If there is a marine pet store in your area, there is a very good chance it has one of these available for sale today. Decorator crabs do not harm other animals in a reef tank. They do not eat macroalgae or hair algae, but they tear off small pieces of vegetation to add to their camouflage, including artificial silk plants. Like arrow crabs, these are normally only available as full-grown specimens.

The existence of decorator crabs as aquarium pets is documented for some of the earliest

Indonesian decorator crab (*Camposcia retusa*), mature specimen

aquaria (Butler 1856, Mellen & Lanier 1935). There are many different "decorator crabs." Most of them, including *Camposcia*, are members of the spider crab Superfamily Majoidea, but there are decorators from unrelated groups seen at pet stores from time to time. Though far less common, one tropical American species, the furcate spider crab (*Stenocionops furcatus*) is seen at marine shops on occasion. It can be collected as far north as the coast of Georgia in the U.S., throughout the Gulf of Mexico, and south to Brazil.

Availability: *Camposcia* from Indonesia are very common at marine pet stores.

Size: Carapace width is 20-24 mm, length 30-36, and the legspan is around 120 mm. Decorator crabs are the largest true crabs seen in reef aquaria. Although they are massive and often sit on the side or top of rocks during the day, specimens can be a challenge to locate among the rocks.

Feeding: This species does not readily acquire food in a reef aquarium. Specimens should be hand-fed pieces of fish or shrimp for long

Indonesian decorator crab (*Camposcia retusa*) covered in anemones and algae from the aquarium. It sets about matching its surroundings whenever moved to a new habitat.

Japanese giant spider crab female on display at an aquarium

Dying animals like the giant spider crab to the right may start dropping legs due to dis-functional autotomy in sickly animals.

term survival, or housed in an individual cage where they will not have to compete for flake or pelleted fish food. They rarely live six months in marine aquariums because they slowly starve, but they can live a few years or more with adequate feeding.

Reproduction: Planktonic larvae.

Cohabitation: Normally kept individually.

Emerald Crab

Mithraculus sculptus (Lamarck, 1818)

In the 1980s and 1990s members of this genus were considered unwanted intruders on live rock in reef tanks. Tiny crabs would arrive inside small holes in live rock and grow over the next year from the size of a pencil eraser to the size of a fiddler crab. Towards the end of the millennium it was discovered this crab is one of the few creatures that eats bubble algae (*Valonia* sp.), a macroalgae considered as a pest in coral culturing. *Mithraculus* (at the time known as *Mithrax*) quickly went from unwanted to beloved.

Specimens are usually only found at night, when they are observed with a flashlight (red LED flashlights are suggested). It is very difficult to find specimens in a reef tank because they hide during the day. I have had dozens over the years and have not seen one live longer than two years, probably because they are full-grown on arrival. *Mithraculus coryphe* is sometimes mixed in with this species at pet shops, but the carapace is diamond-shaped rather than circular. Other related species like *Mithraculus ruber* and *Mithraculus forceps* are less common but rarely confused because they are brown to reddish in color. *Mithraculus* are spider crabs from the Family Majidae.

Availability: This is currently the most commonly sold true crab in the marine aquarium hobby.

Size: 15-20 mm across the carapace.

Sometimes other *Mithraculus* are sold in groups of *M. sculptus*.

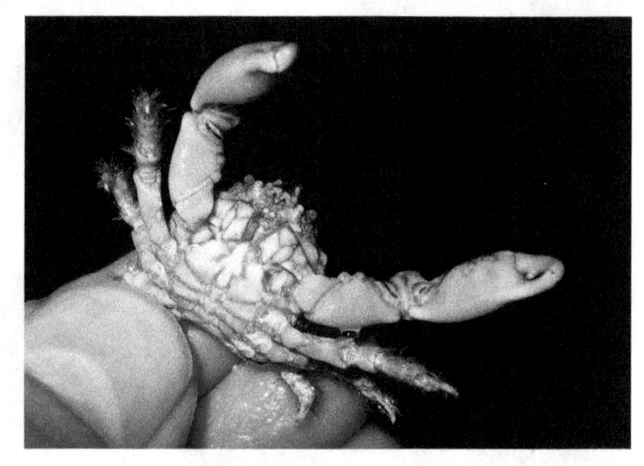

Mithraculus coryphe, male underside

Feeding: They eat most types of fish food and some macroalgae.

Reproduction: Males are the larger of the two genders and have elongated arms; large males can have fantastically oversized front arms, though the claws are a small portion of the total length. Pairs are very rarely seen mating (unrelated to a recent molt). The planktonic larvae have probably never been reared in captivity.

Cohabitation: A number of specimens can be kept in the same aquarium, but males do end up missing a front arm from time to time.

Mithraculus sculptus females tend to be smaller overall and have much smaller chelipeds.

Mithraculus sculptus male, missing a left arm

Hepatus epheliticus male, 81 mm carapace width

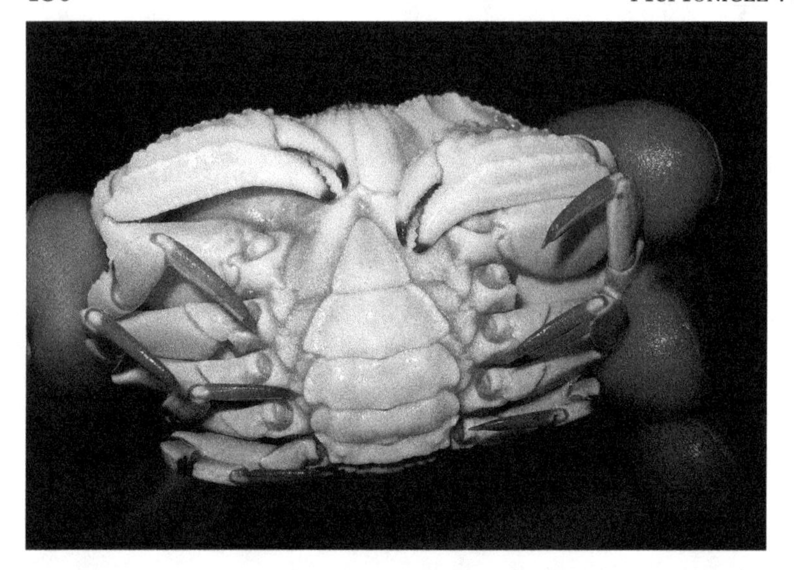

Hepatus epheliticus, female abdomen

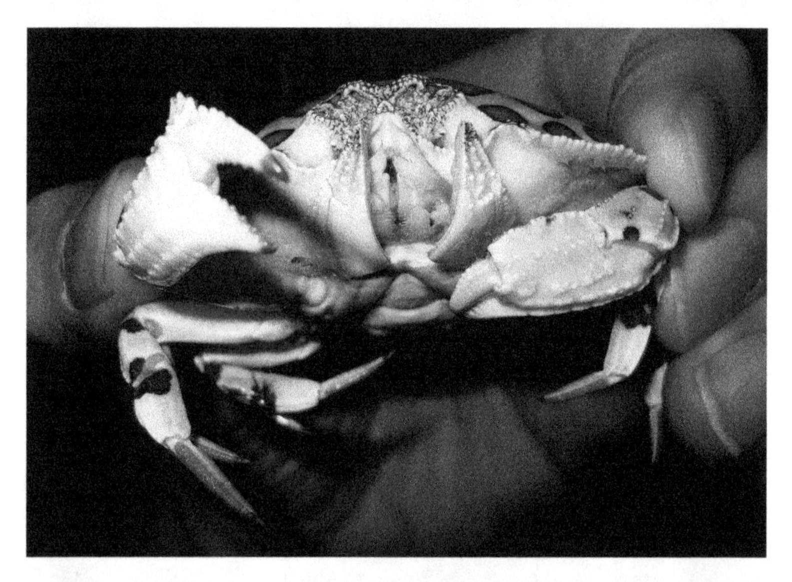

Hepatus epheliticus, female mouth and close-set eyes

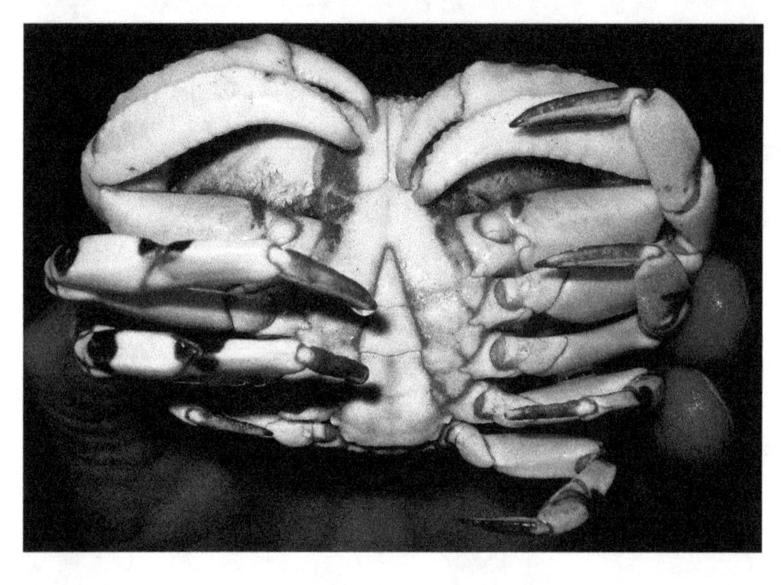

Hepatus epheliticus, male abdomen

Calico Box Crab

Hepatus epheliticus (Linnaeus, 1763)

This is one of the crabs often called box crabs because it folds its legs up over its face (the final product is not square, but the name rock crab was already taken). The species name is Latin for "freckles," though the large-colored spots are massive compared to human freckles. The calico box crab is a cold- and warm-water species found in the Chesapeake Bay, south through the Gulf of Mexico. It is not eaten by humans but it is heavily fished for use as bait for the octopus fishery (Hernáez et al. 2012). This species may be found with *Calliactis* anemones on its carapace, like those found on marine hermit crab shells (Ruppert and Fox 1988). However, I have not yet found a photograph or seen a live crab with one attached. Specimens can live for years in a dedicated marine aquarium. If the aquarium has deep gravel, they bury themselves with just the mouth and eyes sticking out.

Availability: Specimens were commonly seen during the 1980s at local pet shops, but I have not seen one at a shop in decades. Calico crabs are currently available and can be purchased through mail-order from a number of online marine vendors.

Size: To 81 mm (3″+) across the carapace.

Feeding: *Hepatus* eat pieces of clam, mussel, snail, processed fish, and shrimp. Fish food

Calico box crab (USFWS)

flakes and pellets can be too small since available specimens are large. Healthy, live fish are usually safe in the same aquarium, but these crabs should be kept away from slow-moving vertebrates and invertebrates.

Reproduction: Males and females look similar. The male's abdomen is triangular and narrows towards the end, while the female's abdomen is widest towards the middle (both end in a triangular telson). Females produce thousands of multi-stage planktonic larvae.

Cohabitation: Full-grown specimens can be housed together safely at least for short periods.

Atlantic Blue Crab

Callinectes sapidus Rathbun, 1896

This colorful, large crab found along both the east coast of the United States and the Gulf of Mexico can be a responsive and energetic pet. They can learn that the keeper's approach means food, shuffling around in anticipation. "Sapidus" means delicious and this is a very common food item. Specimens can be bought live at markets, often for much less than the standard price of the least expensive fiddler crab at the pet shop. I thought I was the only person who thought these would make an entertaining pet till I was at an Asian market and saw a mother pick out one of the biggest and meanest blue crabs I have ever seen for her three-year-old daughter, who was enamored by the crabs. From her hand signals and body language it was obvious she only picked out the lone crab as a toy for her daughter. I wanted to warn the mother, though I cannot imagine she would let the tiny girl hold it. Unfortunately, I did not speak her language.

Adult specimens require seawater; they may survive in brackish or fresh water outdoors, but low or no salinity will cause the death of mature adults in an aquarium. A specific gravity of 1.015 to 1.022 is suggested. Husbandry difficulties include: 1. Adult specimens are moderately dangerous and the claws can cut human flesh deeply enough to require stitches. 2. They can eat a lot of food which necessitates good filtration. 3. They escape readily and are difficult to contain. 4. Specimens readily tear up their environment, including filtration equipment, hoses, and wires. They smash air stones into little pieces so plastic or wood diffusers should be used if needed.

Availability: Rescued crabs from the seafood store. Blue crabs are not sold as pets but they are commonly available as live seafood. A number of different, large species of crab can be found at markets throughout the year. The largest common food crabs like the deep-water snow crabs (*Chionoecetes opilio*) and king crabs (*Lithodes* and *Paralithodes* spp.) are unlikely to be seen at any market alive. Most grocery stores carry lump crab meat and crab legs, rarely live crabs. Asian markets or fresh seafood grocers in North America often carry a few types of live crabs (mostly *Callinectes, Cancer*, *Chaceon*, and *Metacarcinus*). Specimens sold as food are maintained only to keep them barely living, are chilled with ice, and the majority will die no matter what conditions they are moved to. It is difficult to keep specimens of most species alive for more than a few days because they are already mortally damaged. No matter the species or conditions, females that are holding eggs stress and die easily and are practically impossible to rescue. Two common species of deep-water *Chaceon* would require a refrigeration unit for long-term care if they can be kept alive at all. Blue crabs are usually not kept on ice, have a good chance of survival, and can be kept in room-temperature water.

Size: 120-170 mm.

Feeding: Blue crabs eat almost anything, but primarily other creatures. The common bait used to catch blue crab is raw chicken. Crayfish, cockroaches, and other crabs are eaten with gusto and the loud crunching noises as a blue crab bites apart a hard exoskeleton can be heard

This Blue Crab is begging for more shrimp.

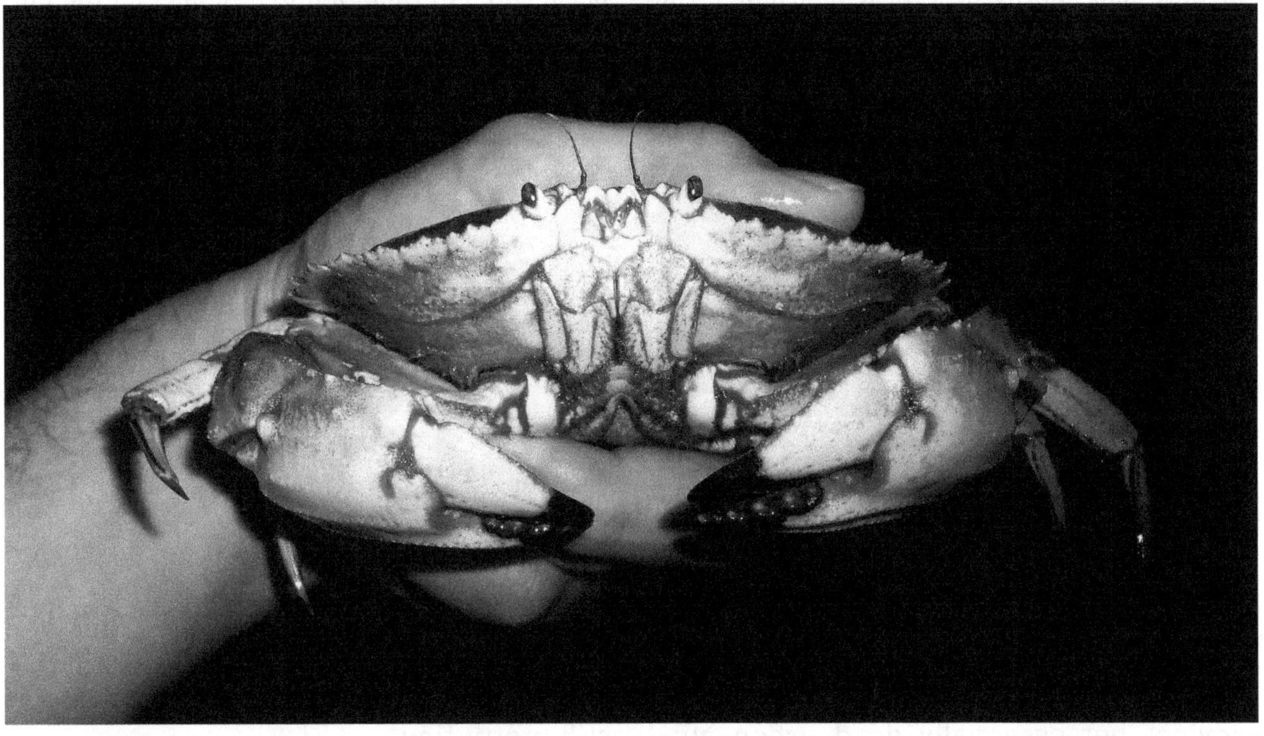

This Jonah Crab (*Cancer borealis*) was kept in a box of ice at al local fish market and barely moved.

across the room. Thawed shrimp tend to float and are quickly located and grabbed from the surface of the water with a flurry of excitement. Blue crabs often reach out of the water and snap their claws together during feeding. Oddly for such a voracious creature, this is the only crab I have kept that does not recognize imitation crab (processed fish) as possible food and refuses to eat it. Still, it eats lump crab meat with gusto and will even chew up carrots, airstones, and imitation lobster (processed fish). Since the crabs recognize my movements as a dinner bell, I can get them to grab other items like peach or pear slices, but they chew on them for a little while and then discard them. The trick with this species is to not overfeed, since no filter can keep up with them. A blue crab has a nearly bottomless stomach for thawed shrimp.

Reproduction: Males and females look similar, except the female has red-tipped claws and a widened abdomen. These classic swimming crabs can only mate shortly after the female molts. The planktonic larvae have been reared in laboratories with wild-collected, strained plankton.

Cohabitation: Multiple specimens can be kept in the same aquarium only as long as they are similar in size and do not molt. If a single crab is kept in an enclosure, it is less likely to try to escape.

Speckled Swimming Crab

Arenaeus cribrarius (Lamarck, 1818)

Overall this species is very similar to the blue crab, just smaller. Young specimens should be kept on a layer of fine sand so they can hide. They often remain submerged with just the eyes and mouth visible during the day. They are able to bury themselves in the blink of an eye when disturbed. The sand should be less than half an inch (~10 mm) to prevent anaerobic bacteria growth. Collected specimens are often only 20-30 mm across the carapace (spine to spine). Of course, the cute little guy brought home can grow rather large over time. A small specimen molts every six to eight weeks and grows notably. It will not be so cute after half a dozen molts and the pincers that could barely give a silly little pinch will eventually be able to slice flesh.

This deep sea red crab (*Chaceon quinquidens*) from a New England fishery was moving, but essentially dead, when purchased from an fish market.

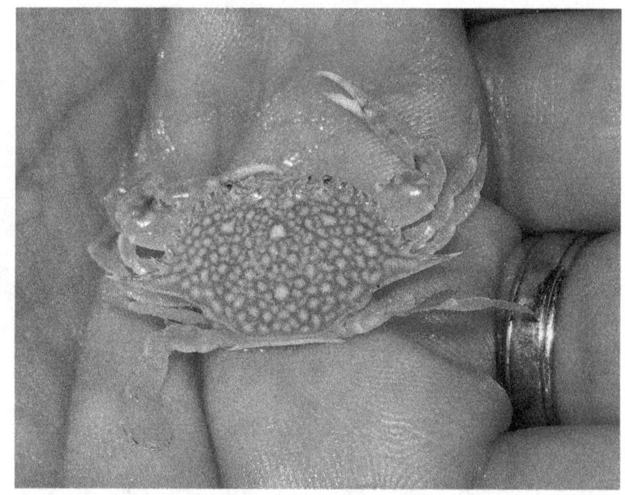

Speckled swimming crab- After one molt in captivity this crab has grown a bit, but it is still pretty tiny.

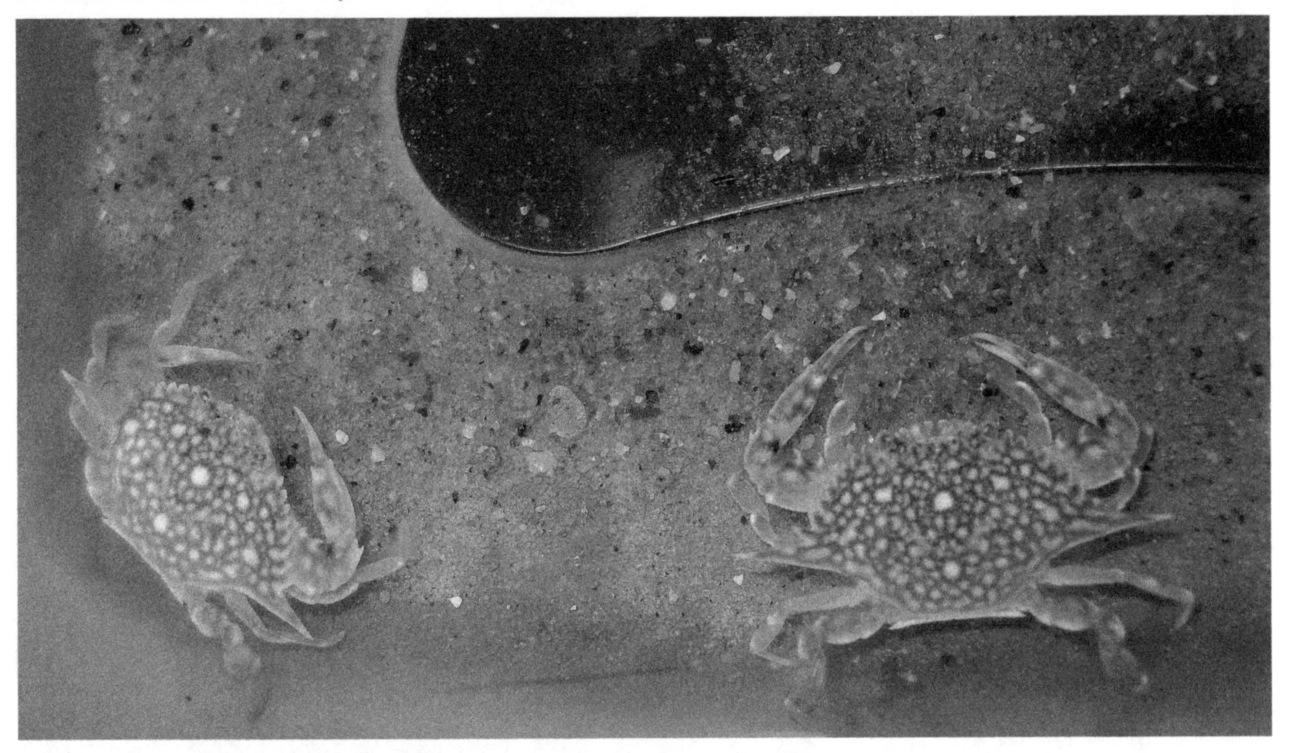

Speckled swimming crabs (*Arenaeus cribrarius*) from Topsail, North Carolina

Speckled crab: after two molts the pinch from the larger claws begins to hurt a tiny bit.

Availability: I have never seen a species of swimming crab at a marine pet shop, probably because they require dedicated housing and would decimate a reef or fish tank. However, these can make interesting pets that live for many years. Although *A. cribarius* are not offered for sale in the U.S., it is easy to find small specimens in the water off many beaches along the east coast of the United States. This species is found from Massachusetts south to Uruguay (a massive stretch of coast 42° north of the equator to 33° south). This is a popular food item in Brazil where adults are available at markets.

Size: They mature at around 60 cm, but can ultimately reach 100-150 mm in nature (Pinheiro and Fransozo).

Feeding: Eats fish flakes and pellets, any seafood, and a huge variety of animal and vegetable matter (though vegetable matter is not suggested and is taken only if they are extremely hungry). Specimens commonly dance around the

Purple claw hermit crab, Saint Kitts © Jason Ross

Coenobita sp., Okinawa © Richard Cook

substrate when food is introduced. Large specimens of *A. cribrarius* have been documented to feed on baby loggerhead turtles (Frick 2003).

Reproduction: The planktonic larvae are not easily reared in captivity.

Cohabitation: It is possible to keep multiple specimens together of similar size only until one molts. Freshly molted animals, if not eaten entirely, will have one front claw, and two or three of the legs behind it, eaten off when kept with other specimens. Except for a mature male and female during courtship, this species should be treated as solitary.

SELECT ANOMURAN CRABS

Purple Claw Hermit Crab

Coenobita clypeatus (Fabricius, 1787)

My first experience with this widespread Caribbean species was a pet crab that would never leave his shell. I was probably seven or eight. Eventually I could not help but put my finger in the pincer—at least that would move a little. I still remember the big, purple blister and a bit of pain. I do not know what became of the purple claw that pinched me as a child, but it certainly lived no more than a year or two, when it might have lived decades longer with proper care. With adequate care, *Coenobita* species may live longer than the family dog. Longevity data is anecdotal, but ten to forty years is the common range for a wild-caught specimen provided an adequate captive environment. Fortunately, in recent years especially, the requirements for long-term care have been better described and publicized.

Terrestrial hermits are fantastic survivors capable of staying alive after weeks without water and longer without food. Their flattened legs allow them to seal the entry of the shell to retain moisture. Hermit crabs can hole up in their shells to protect from moisture loss, but most pet crabs die because they do not get enough water. Even if water is available they will still eventually dry out if the humidity is too low. All *Coenobita* should have regular access to a dish of salt water, a freshwater dish, a food dish, and deep substrate to construct molting chambers. These items are not difficult to provide, though half a foot (15 cm) of substrate eliminates certain cages or makes them too easy to escape from. The water bowls should be deep enough for the largest hermit to submerge its shell halfway, but remember that terrestrial crabs can drown. The bowls should have rocks or other climbing items to make it easy for the smallest hermit to escape.

For many beginners, saltwater is the least accessible item, but department stores now sell small bottles of artificial seawater for hermits. Over a long period, it is much cheaper to mix your own salt. The chosen salt should be an artificial seawater mix, not the salt used to medicate freshwater fish. Follow instructions on the box, but keep in mind it is safer to make a weak mix than an overly strong one. Artificial mixes normally contain a dechlorinator, so it is not necessary to aerate the water for a few days to remove the free chlorine used as a disinfectant in many potable water supplies.

Habitat Pests: For a number of years when my children were young, we traveled to Virginia Beach to visit family and enjoy the sand and waves. The tourist shops had big displays of hermit crabs that were hard to ignore. The first year my youngest daughter convinced my wife to buy her one. Six months later, a neighbor six houses down the street brought a healthy little hermit to my front door because I was the neighborhood "bug guy." I only discovered it was Gwyn's hermit after I learned my wife had dumped the cage in the backyard weeks earlier (some small maggots appeared in a water-logged food dish and grossed out the girls). The maggots were phorid fly larvae that had capitalized on a food dish full

While color can be variable and unreliable in identification, the shape of the large claw of the Ecuadorian (left) and purple claw (right) are quite different.

Coenobita brevimanus in Panay, Philippines. This is the largest of the *Coenobita* with adults weighing up to half a pound (230 g) (Fox 2000). It is commonly traded as the Indo hermit crab. Specimens are seasonally available from specialty online vendors. © Christian Schwarz

Coenobita in enclosure.

of wet hermit crab pellets. Infestation of the food dish is a common concern with hermits, because overfeeding is easy and the food often sits in the dish for weeks. Phorid flies are found everywhere, but maggots are an uncommon problem because the food dish is usually kept dry.

There are many possible pests. One of the most common pests encountered with pelleted food or dried krill kept in the dish too long is an infestation of grain mites. They can tolerate extreme dryness if the humidity is high. Following a population explosion, grain mites have a phoretic stage which attaches to the crab. These look like tiny white sand grains stuck to the animal, often on flat areas near segment joints. They can be manually removed with some effort, but they are mobile and return quickly if not destroyed. Hermit crabs can be harmed during manual removal and it is nearly impossible to get every mite. The best way to reduce or eliminate grain mites is to reduce feeding to what the crab will finish in a day and to add clean-up crews to eat scraps that get mixed into the substrate. The most commonly employed crews are springtails and very small isopods. However, it is important to remember the clean-up crews cannot survive low moisture as long as the mites can. Dry food kept in the dish for long periods can become infested with meal moths, dermestid beetles, or drug store beetles which require no moisture and very little humidity. These stored product pests are reduced or eliminated by removing and replacing uneaten food in the dish at least once a week.

Grain mites can be confused with mites that attach to the legs or modified gills of wild-caught hermits (it is unlikely to ever see a captive-bred hermit). Harmful mites attach to the joints or abdomen where the exoskeleton is easy to pierce for feeding. Parasitic mites will not be found eating food scraps and rarely are unattached. These mites may be reduced by placing the crab in alternating saltwater and freshwater baths. The shell is placed upside-down so the water runs in. As soon as the crab opens up to leave, it is placed in the freshwater the same way. There should be an easy way for the crab to escape the water or the keeper must not walk away.

SHELLS: It is important to have some empty shells available in the habitat, especially if there are multiple specimens in one enclosure. *Coenobita clypeatus* prefer thick, heavy shells that they can retract entirely within so that just the one big claw covers most of the entrance. Other hermits prefer smaller shells or the shells of specific snail genera. Shells can be purchased from craft stores and pet stores anywhere, or from sea shell stores in coastal tourist areas. Shells should be boiled to remove pesticides or parts of old inhabitants before they are added to the enclosure. Painted shells are generally frowned on because the paint could chip off and be consumed. Hopefully poor husbandry is not blamed on paint chips. Heavy shells are normally chosen for protection. It would seem bigger, lighter shells might be chosen, but I have never been able to get one to pick a freshwater snail shell.

Philippine coenobatids are regularly found very far away from the shore and use the shells of large land snails. They may never return to the shore to release larvae, but possibly release them in rivers where they flow to the sea. Also, due to the lack of empty shells, they often predate on the initial inhabitants in order to obtain the snail shell. They also need new shells when the old ones fade to white and are no longer cryptic (pers. obs. Schwarz).

EXOSKELETON REPLACEMENT: In the first nine months of life hermit crabs can molt more than a dozen times, mostly in the ocean. The initial molts on dry land are probably exposed, similar to marine hermits. However, later molts grow further apart and require a subterranean molting chamber. Terrestrial hermits (unlike marine species) must construct underground chambers for protection during the molt. They can refuse to molt for very long periods without a chamber, especially when kept with other specimens. In captivity, large hermit crabs should molt about once a year, and smaller animals as often as twice a year if good conditions are maintained. Molting in captivity requires an undisturbed area. A solitary specimen may eventually molt out in the open but the molt will be delayed. If multiple specimens are kept together and there is no place to molt, they can hold off molting for years.

The keeper should provide a compactable substrate for the hermit to build a molting cell. Five to six inches (127-152 mm) of damp sand should cover the bottom of the terrarium. Shallower substrate (deeper than the host shell's width) can be used if there are flat rocks for molting chambers to be constructed under. If the sand becomes dry, the chamber will collapse and the crab may die or be killed by tank mates. If the sand is too wet, the lack of oxygen or toxic gases produced by anaerobic bacteria can kill a crab in its molting cell.

Play sand or sand bags are easily acquired and the most commonly used for substrate. Both contain silica-based sand. The material can be washed to remove dust, but washing is usually unnecessary and a third of the sand washes away as fines. If the sand smells strongly or appears oily, a new source should be sought—washing is not likely to be enough. Calci-sand made of calcium carbonate should not be used since it hardens after moistened and dried. It can imprison a hermit. It is also chemically reactive and could damage or block up the digestive system of invertebrates when ingested with food. Peat or coconut fiber are often mixed into silica sand to help indicate when water is needed since these materials change color as they dry. Dirt or clay could be used instead of or in addition to sand, but these substances build up on the animals' legs, eyes, and antennae. Attached dirt is usually a cosmetic issue, but it can build up enough to cause stress and physical damage.

The molt itself usually takes less than an hour, but the crab spends about thirty days in the molting cell. After building the cell it sits within the chamber for about two weeks while the old exoskeleton is partly absorbed from the inside. After the molt another two weeks are needed for the new exoskeleton to harden.

Availability: *C. clypeatus* is possibly the most widely kept pet invertebrate in the world, certainly the most common in the New World.

Size: It is difficult to measure a hermit crab. It is nearly impossible to take a measurement of the width or length of the carapace from a live specimen. Even when the measurement is taken, it may be difficult to estimate the overall size from a carapace width of 20 mm (though that is a rather big hermit when the carapace width is close to 40 mm). Commonly available *C. clypeatus* (including the host shell) are about the size of a golf ball, but full-grown specimens are closer to the size of a baseball.

Feeding: Fish meal pellets and dried krill are the commonly available foods, but hermits ignore these foods for long periods. The ingredients of pelleted food should be checked for preservatives, specifically ethoxyquin, since it can result in deformities or death during molts. Live insects and fresh seafood (thawed shrimp) are often eaten within hours. Stronger smelling fruits like sliced banana, mango, and strawberry may be eaten quickly, whereas apple, pear, grape, and peach often go unnoticed. Vegetables including corn, sweet potato, and carrots will be consumed eventually if the crabs are hungry enough. Dried hardwood leaves are commonly nibbled on. Coconut, various seeds, and nuts may be eaten. However, many nuts, including walnuts, Brazil nuts, and sunflower seeds, are a good source of copper and might cause death over time if offered as the primary food. While

Coenobita cavipes in Panay. © Christian Schwarz

Ecuadorian hermit crab (*Coenobita compressus*)

Coenobita cavipes has taken over residence from a recently living land snail. The crab's old, faded house and the remains of the snail carcass can be seen on the right. This was found in an upland forest in the Central Panay Mountain Range. © Christian Schwarz

the appetite of marine hermits and aquatic true crabs is minimally impacted by temperature, *Coenobita* species display reduced feeding interest if the daytime temperature does not reach 78-85° F (26-29° C).

REPRODUCTION: Males and females are difficult to tell apart when they are alive. On dead specimens, it is easy to check the base of the rear walking legs for small holes (opening to the gonopores) which only the female has. The multi-stage planktonic larvae of this species start out very tiny and likely could not be reared without an aquaculture facility and strained wild plankton.

COHABITATION: Other than extremely tiny creatures, very few creatures can be safely kept in a cage with terrestrial hermits. These tough scavengers are fierce predators at heart. I placed one small *C. clypeatus* in a cage with some giant desert skunk beetles, *Eleodes longipes*, for short-term housing. I thought the beetles would be protected by their defensive chemicals, ability to run, and hard shells, but the lone purple claw caught and chewed up five of the large beetles in a single night.

Hermit crabs are gregarious in nature and keepers often think of them as social. This is true, but the social graces include sparring, establishing pecking orders through fighting, and of course killing. *Coenobita clypeatus* usually cannot hurt each other, but animals that have molted in the last few days will be pulled from their shells and killed by cagemates if accidentally unearthed by the keeper (or if the molting cell collapses). Many keepers maintain their crabs in large colonies and feel the social interactions lead to more active and less skittish animals.

Ecuadorian Hermit Crab

Coenobita compressus H. Milne-Edwards, 1836

Specimens range in color from brown to steely blue, and their big claw is usually the same color as the rest of the body. The shells they choose seem too small for their bodies. They modify the shells over time to fit better and tend to pick the same shell after they molt unless it is severely undersized. Ecuadorian hermits are better climbers and far better escape artists than other common hermits. Also, they generally only live two to five years in captivity, but this is likely because they are already fully grown when acquired. A small specimen might be as long-lived as *C. clypeatus*.

AVAILABILITY: This hermit has been the second most common *Coenobita* available in North American pet shops for decades.

SIZE: A full-grown specimen is around 23 mm in carapace length and 14 mm carapace width. Available specimens are full-grown, so they tend to be a little bigger than the "small" purple claws ("small" being the most common size available for sale, between the size of a ping-pong ball and golf ball).

FEEDING: Fish food pellets, krill, live insects, certain fruits, dead leaves, etc.

REPRODUCTION: Multi-stage planktonic larvae.

COHABITATION: They communicate by stridulating, but what exactly they are saying is uncertain. It could be, "Get out of my territory." Like other *Coenobita* they seem to be gregarious but fight over resources, including shells and food. They also kill molting animals when exposed.

Strawberry Hermit Crab

Coenobita perlatus H. Milne-Edwards, 1837

Strawberry hermits are often sold full-grown (many times larger than a small purple claw) and seldom live past a few years. A young, small specimen would likely live a lot longer, but a larger size may be a factor in surviving the long trip from Asia. They are deep reddish overall and the bumps on the red-colored legs are often pale, almost white in color. These small white bumps on red give the vague impression of the seeds on a strawberry's flesh. Wild-collected specimens

This wild *Coenobita perlatus* was imported with part of a leg missing. Hermits do not readily lose or regenerate legs. This specimen has been in captivity for a year and has not yet molted. Still, it is unlikely the leg will grow back correctly even over multiple, successful molts.

Strawberry hermit crab (*Coenobita perlatus*)

Blueband hermit crab, *Pagurus samuelis*, California (Jerry Kirkhart)

often have difficult molts in captivity, while survivors can end up very pale if the food does not have the natural levels of carotene (similar to the way a flamingo loses color in new feathers).

Strawberry hermits are more active than the average purple claw. However, they may become lethargic if daytime temperatures fall below 78° F (26° C). Like the Ecuadorian, they often pick small shells they cannot fully retract into, usually the very heavy shells of large *Turbo* spp. snails.

Availability: Often seasonally available from online vendors and from some pet shops. They are usually imported from Indonesia or Malaysia, though they are found widely across the Indo-Pacific. This is the third most common species available in the North American pet trade. Though it was pictured in older books (Walls 1982), this species was almost impossible to find until around five years ago. Availability of Old World species has improved with increased interest in recent years.

Size: Strawberry hermits seem to be a little smaller than full-grown *C. clypeatus* because they choose smaller shells. The carapace length of a large, older adult reaches about 40 mm.

Feeding: Fish food pellets, krill, live insects, dead leaves, certain fruits including strawberry and mango, and vegetables such as carrot, corn, and sweet potato.

Reproduction: Multi-stage planktonic larvae.

Cohabitation: They seem to be more solitary than the previous species, as specimens tend to choose individual hiding areas if given a large terrarium and multiple hides.

Australian Hermit Crab

Coenobita variabilis McCulloch, 1909

This hermit looks a lot like the Ecuadorian, but grows larger. Like most species it has a limited range of common colors, however, unusually pale, dark, or unusually colored specimens may be found.

Availability: This species is commonly kept in Australia but it is only rarely available from specialty vendors in North America.

Size: Approximately the same size as *C. perlatus*.

Feeding: Fish food pellets, krill, live insects, certain fruits, dead leaves, etc.

Reproduction: Aussie hermits have relatively large eggs and larvae so it is possible for them to be reared in captivity by the dedicated hobbyist. The larvae hatch out at 4-5 mm and are big enough to eat baby brine shrimp. As with any planktonic crab, even if successful, the survival rate tends to be dismal and water quality must be well-maintained for even a few to survive.

Cohabitation: Commonly kept in groups.

Coconut Crab

Birgus latro (Linnaeus, 1767)

This monster, like our land hermits, is a member of the Family Coenobitidae and is more closely related to them than to any of the marine hermits. *Birgus* dig burrows like most of the common terrestrial hermit crabs and may amass a trove of shiny objects in their dens. Therefore, they are known as robber crabs. Coconut crabs are often tan or brown in color, but they can be quite beautiful. Some specimens are brightly marked with yellow, red, purple, or blue. This species is too big and strong to be kept as a pet, since a large, reinforced cage, unlike anything on the market, would need to be built to safely house and contain an older adult crab. While the claws are probably not strong enough to remove a finger, they could cut to the bone. This species is severely threatened by hunting.

Availability: Small specimens might be mistaken for *Coenobita* spp. because they live inside abandoned snail shells for the first two years of life. However, the idea of acquiring a coconut crab by accident is an urban legend. It is possible for a small specimen to be mixed in with tropical

Birgue voleur (Birgus latro. *Fabr.*)

Asian hermits (*Birgus* is an Old World genus and could never be accidentally collected with New World species). Nevertheless, any hermit for sale in a shell is unlikely to be *Birgus*. I have only seen a few pictures of this species being kept as a pet in Japan and the large size of the specimens suggests they were bought from a local fish market where they were being sold as food. (This species is an apparently delicious and expensive meal.) Coconut crabs have unfortunately seen limited availability in the pet trade in Germany (Zompro and Fritzsche 2008). Specimens the size of a man's fist were often available at reptile shows in the United States during the 1970s (pers. comm. Kreuger).

Size: Although small compared to some of the largest marine invertebrates, these are by far the largest and most massive terrestrial arthropod. The body length may reach 400 mm while females are usually capable of reproduction at 40 mm (Amesbury 1980). The legspan can reach a meter.

Feeding: Anything a standard hermit would eat, plus coconuts and small vertebrates.

Reproduction: Since it has no shell this is the only coenobitid that could be easy checked for gonopores on a live specimen. Multi-stage planktonic larvae.

Cohabitation: Specimens kept in communal pens do not kill each other, but they will not molt (Amesbury 1980).

Saltwater Hermit Crabs

There is a fantastic variety of saltwater hermits, some of which may be more closely related to other anomurans than to terrestrial hermits. Most hermits use a specific range of abandoned snail shells to protect their soft abdomens. Some species accept a narrow range of host shells and some are adapted only for elongate shell openings. There are a few atypical species that form "shells" by drilling into sandstone, grow their own living shells from sponges or hydrozoans, or carve homes from bits of bamboo. Some hermits place stinging anemones on their shells for added protection from predators (specifically from the octopus). Some filter-feeding hermits encase themselves in living corals and collect plankton from the passing water like a feather duster worm—these are very difficult to keep alive for long. Many saltwater hermits are easy to maintain and can live a decade in captivity, even the small ones.

The most popularly kept species are the clean-up crew hermits used to control filamentous hair algae (*Bryopsis* spp.) in reef tanks. These include:

Blue Leg Hermit, *Clibanarius tricolor* (Gibbes, 1850). These Caribbean hermits were the first popular clean-up crew species and they have been regularly available for a few decades. Available specimens are usually tiny, and come in elongate shells that are often less than 10 mm. I have two blue leg hermits that are seven years old and are larger than a full-grown scarlet or zebra hermit. One carries a full-grown *Astrea tecta* shell and the other a medium-size Turbo shell. Most do not live so long or grow so large; they had countless counterparts and predecessors who are no longer around. Large specimens have a carapace 14 mm long and 8 mm wide at the base, but they can reproduce at 6 mm carapace length.

Red Leg Hermit, *Clibanarius digueti* Bouvier, 1898, from the Gulf of Mexico are usually available close to adult size. They prefer to carry elongate shells that are rarely greater than 3 cm. They grow very little, even after three years in captivity. This is the smallest (at full size) of the commonly available species. Since red leg hermits are sold as mature adults, the available specimens are much larger than the more commonly available immature *C. tricolor*. Mature specimens have a carapace 6 mm long and 4

Blue leg hermit with the rear legs visible as it tries to dislodge fingers from its shell. This specimen has been in captivity for seven years and is far larger than available specimens.

Large, old blue leg hermits (*Clibanarius tricolor*)

Red leg hermits (*Clibanarius digueti*) in their original shells. They remained in these shells for two years.

Zebra hermits can be long-term inhabitants of anemone tanks.

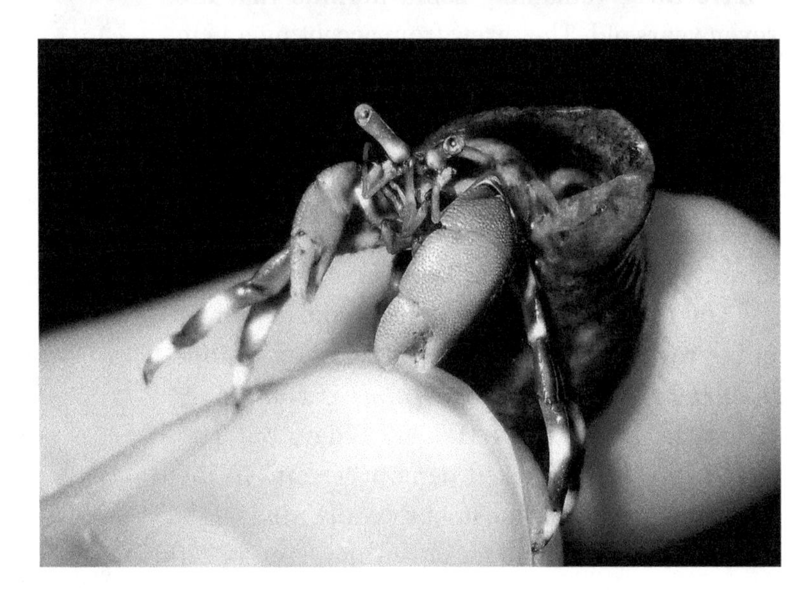

Zebra hermits are small at maturity, but are usually sold as tiny immatures, a fraction of the adult size. This seven-year-old specimen is as big as they get.

Zebra hermit (*Calcinus seurati*) next to a large blue leg hermit

mm wide at the base, but a large older specimen sometimes obtains a carapace length of 10 mm.

Scarlet Hermit, *Paguristes cadenati* Forest, 1954, is Caribbean in origin and was among the first clean-up crew hermits in the early 1990s. I have three scarlet hermits that are ten years old and still doing well (I started with four). They are normally sold at full size (8-12 mm carapace length) and molt once or twice a year without getting any bigger.

Zebra Hermit, *Calcinus seurati* Forest 1951, from Hawaii was commonly available at pet shops around 2010 and is still available by mail order. I have three remaining zebra hermits that are seven years old. They grew from accepting a 1 cm (elongate) shell to a 2.5 cm shell. Their carapaces are 6 mm wide by 9 mm long, a little smaller than mature scarlet hermits. The left claw is greatly enlarged and is used to cover the entrance of the shell like terrestrial hermits (where most marine hermits have roughly similar-sized chelipeds).

The Indo-Pacific *Ciliopagurus* spp. are large, fully-grown flat hermits sold as "reef-safe." *Ciliopagurus strigatus* (Herbst, 1804) is commonly called the Halloween Hermit today, but was called the "white-backed demon hermit" in the early days of the marine hobby (Walls 1982). It is sometimes sold online for clean-up crews, but it

Scarlet hermit (*Paguristes cadenati*) and a zebra hermit, very old specimens

Myrtle Beach tide pool, where white hermits can be found © Ryan McMonigle

Red leg hermit, older adult in a large shell.

is a scavenger. It is more likely to eat other tank inhabitant than hair algae.

Pagurus longicarpus are often sold as White Hermits for clean-up crews because they are easily collected from some East Coast beeches. Specimens of *P. longicarpus* are very active and feed well. However, they rarely live six months in a reef tank because they are not adapted for continuous submergence.

White hermits (*Pagurus longicarpus*) from a sandy tidepool in Myrtle Beach, South Carolina

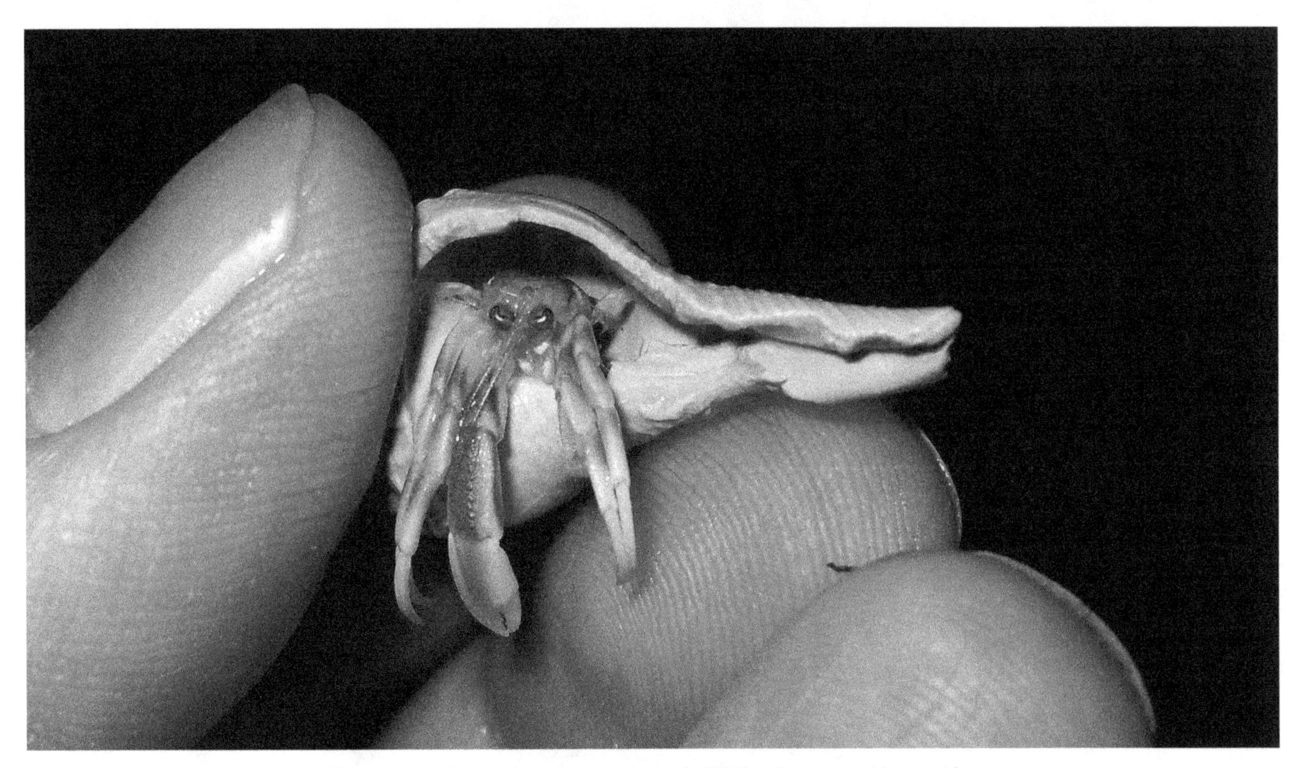

Pagurus longicarpus are a little larger than the maximum size of more common clean-up crew species.

Blackeyed hermit crab, *Pagurus armatus* (Ed Blerman)

Hairy yellow hermit crab, *Aniculus maximus* (Ratha Grimes)

Large marine hermit (NOAA)

Anemone hermit crabs (Bernard Dupont)

Small hermit with anemone (NOAA Ocean Explorer)

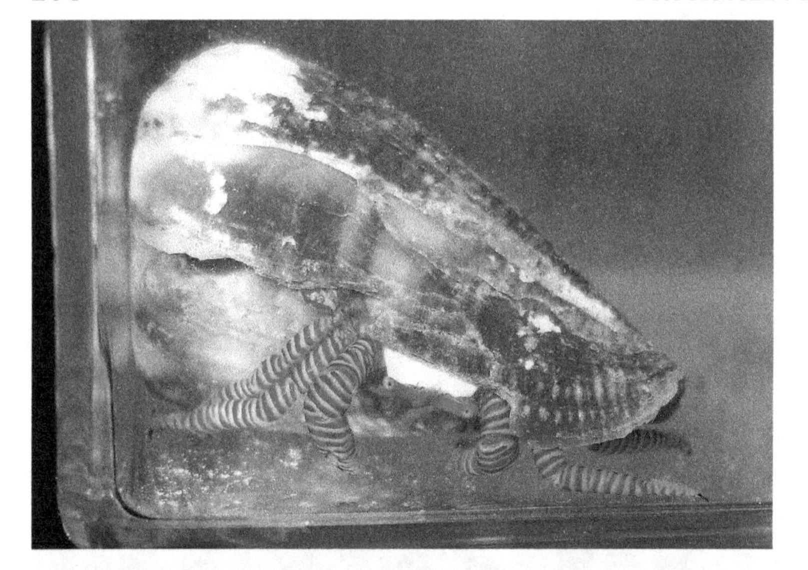

Ciliopagurus strigatus and its relatives live in cone snail shells that have a wide, flattened opening so the crab's body seems two-dimensional compared to other hermits. This and another Halloween hermit proceeded to cannibalize a recently molted tank mate a few minutes after the photo was taken.

Dardanus calidus from Florida

Dardanus calidus from Florida

The following two hermits are commonly seen on display:

Thinstripe Hermits, *Clibanarius vittatus* (Bosc, 1802), are found from Virginia to Brazil and can be large enough to carry a 4″ (10 cm) shell. I caught a big specimen in a 3″ (7.6 cm) shell with a crab pot while casting for blue crabs in Virginia Beach. This is the common species used in "touch tank" displays at aquariums. Medium-sized specimens with a shell about the size of a ping-pong ball are often available at marine stores, but they are too large to be safe for a reef tank. I have maintained one in a reef tank for six years since it came mixed in with a group of tiny blue legs and has not grown overly large. Over four years it grew big enough to hold a *Turbo fluctuosa* shell and has lived two years at that size.

Anemone Hermits (*Dardanus* spp.) are commonly available with anemones on their shells. These are available in shells smaller than a golf ball. Sometimes huge specimens in big whelk or helmet snail shells are seen. Huge specimens require a lot of food and are destructive.

Availability: Small species used as clean-up crews are available year-round from marine pet stores and online vendors. Larger species including those with anemones on their shells (usually *Dardanus* spp.) are encountered seasonally at local pet stores.

Size: Clean-up crew species often have a carapace length of a few millimeters and a width of around 1 mm. Most have carapaces that never exceed a centimeter long. Large *Dardanus* specimens have carapace lengths exceeding 40 mm.

Feeding: Every species will eat fish flakes and pellets and most will eat fragile macroalgae (*Bryopsis*, *Ulva,* and *Enteromorpha*), though not *Valonia*, *Caulerpa*, etc. Large specimens (any crab in a shell bigger than an *Astrea tecta* snail) are capable of eating most other marine invertebrates. However, anemones and corals are not considered food and most fish easily avoid predation. Large specimens will eat the legs off serpent stars and even chew the spines off sea urchins. Oversized specimens should be hand-fed shrimp and processed fish or they will eventually suffer a bad molt from inadequate nutrition.

Reproduction: Multi-stage planktonic larvae. They hatch out much larger than marine true crabs and look like tiny (1.5 mm) shrimp. I have had female blue leg hermits release young in the aquarium once a month for up to six months in a row.

Cohabitation: Similar-sized specimens of various genera and species can be kept together with relative safety though they do fight over food and shells and sometimes eat smaller specimens. Different genera and species of similar-sized specimens cohabitate, but very large hermits, specifically those from the genus *Dardanus*, may eat smaller species. Wrasses, arrow crabs, and some other common saltwater tank inhabitants cannot be kept with clean-up crews since they will eat every little hermit.

Porcelain Crabs

Porcelain crabs are easily differentiated from true crabs because they have long, thick second antennae and only six walking legs. They also have large compound eyes, but the eyes are fixed. Like most anomurans, they have greatly reduced rear legs. The common name comes from the pale exoskeleton of the type genus *Porcelliana* that resembles white porcelain (usually with areas marked with red polka dots). Most species are associated with anemones or corals, but do not require them in captivity. These live six months to two years in a reef tank.

Availability: Members of the Family Porcellanidae have been common at marine pet shops since at least the 1980s.

Size: 12-15 mm across the carapace.

Feeding: Porcelain crabs have been popular because they are "reef safe" and feed on plankton rather than tank mates. The maxillipeds are

Aliaporcellana sp., Porcellanidae (Christian Gloom)

Porcelain crab in anemone (Lakshmi Sawitri)

Porcelain crab pair, the slightly larger male is on the right but the genders are difficult to discern from above.

adapted into large, fan-like baskets that are opened and closed to strain planktonic food from the water. However, plankton is not necessary, as they use their chelipeds to scoop flake or pelleted fish food that falls to the bottom into the maxillipeds.

Reproduction: Planktonic larvae have probably never been reared in captivity.

Cohabitation: Pairs usually hang out, almost touching, but otherwise are solitary. They do not eat each other but can damage one another if there is not enough room to set up territories.

Porcelain crab: note the antennae are outside the eyes and there are only three pairs of walking legs.

Mole crab legs face both directions (the eyes and mouthparts are on the left).

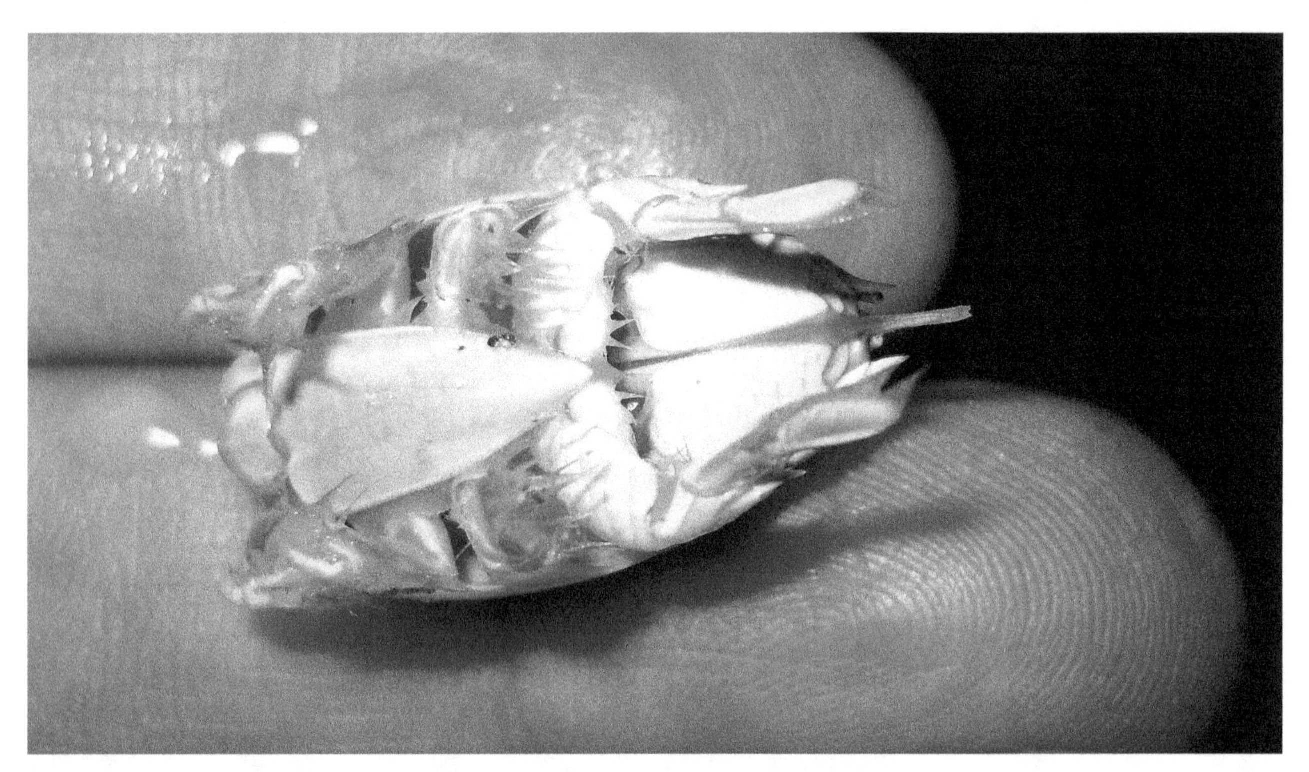

The mole crab abdomen ends in a long, narrow, terminal segment that aids in burrowing.

After burrowing, the eyes and antennae stick out of the sand.

It can be difficult to discern the front and back of a mole crab because they swim backwards and the tiny eyes are at the end of long, inconspicuous stalks.

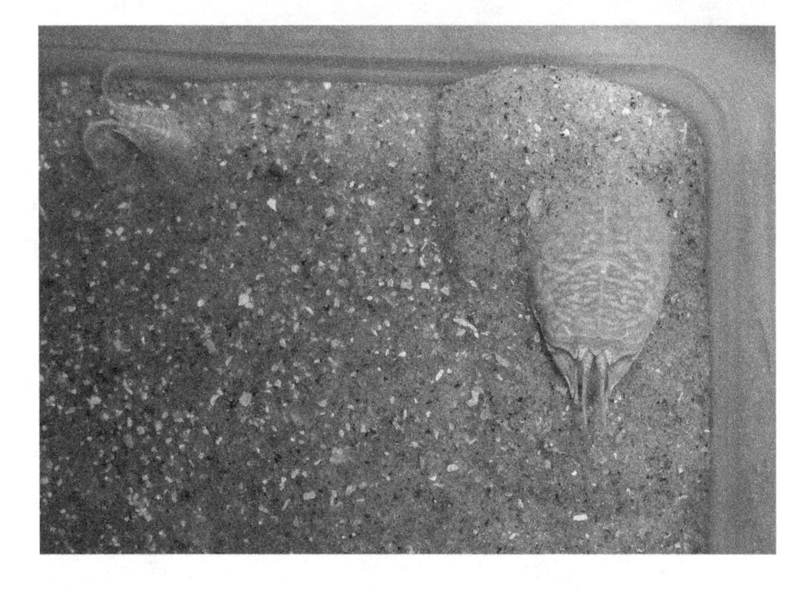

Filter feeding: the small mole crab on the upper left has extended the large feathery mouthparts used to collect plankton.

SELECT XIPHOSURAN

Atlantic Horseshoe Crab

Limulus polyphemus (Linnaeus, 1758)

The horseshoe crab is also known as the king crab in many areas, but it is not a crustacean. This ancient creature is very different from brachyuran and anomuran crabs since it has a unique body form, visible book gills, and every leg ends in a chela. The small chelae at the end of each leg are used for locomotion and prey capture. They do not have defensive pincers like crabs. The book gills and legs are often used for swimming, especially for small specimens to escape. The long telson is another stand-out feature. Some people confuse the telson with the tail of a stingray, but the horseshoe crab is harmless to humans. The telson is used to right the animal when it gets flipped on its back.

This is one of the most common marine arthropods kept as an aquarium pet. This species requires close to full salt (~1.020 salinity) though a different, rarely available species from tropical Asia, the mangrove horseshoe crab (*Carcinoscorpius rotundicauda*), can be kept in brackish water. The horseshoe crab body is not built for climbing, so escape from a standard aquarium is highly unlikely. The enclosure should include an outside power filter (a sponge filter is not enough to process wastes as the animals grow), a shallow layer of sand, but no rocks or decorations. Specimens are easy to keep alive for years if there are no rocks in the enclosure. They occur in mud and sand flats where rock outcroppings are rare. In captivity they wedge themselves under overhanging rocks when they are about to molt. This almost always results in molting damage and death.

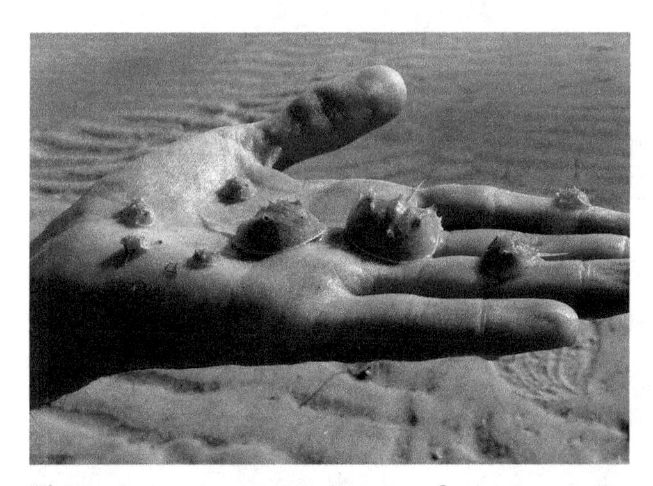

These immature specimens of varying ages were found in shallow water off Lover's Key, Florida (during kayaking 5-4-2008). © Erin Russell

The larger specimen at left has molted once after hatching; this is the first stage capable of feeding. The tail (telson) starts out very short.

Specimens molt half a dozen times in the first year of life and each molt becomes progressively more staggered. Once they are two to three years old, the molts slow down to twice a year. Each molt is accompanied by a notable increase in size. Maturity takes at least six to seven years in the home aquarium, but they do not usually live that long in captivity. I have had specimens live five years that only reached 3.5" (9 cm) in carapace width. Wild specimens reach the size of a serving plate, or larger.

Availability: Most marine and general pet shops carry horseshoe crabs from time to time. This species has been kept in marine aquariums for a very long time (Mellen and Lanier 1935).

Size: Adults can exceed 300 mm in carapace width. Commonly available specimens are usually 30-40 mm across the carapace.

Feeding: *Limulus* eat most pelleted fish foods, pieces of fish or shrimp, and earthworms. They are not predatory on large creatures, but will eat small worms and scuds. These are heavy eaters so bi-weekly water changes are suggested.

Reproduction: Egg masses collected from the shoreline hatch into large immatures with direct development. There are no swimming planktonic stages. They do not eat before the first molt and are about 4 mm across at second instar. They are easy to raise up in captivity, but there is some die off in the early stages. Unfortunately, getting adults to mature and spawn in captivity would be a serious challenge of time and space. The need for large caging is one problem, but an artificial beach shore would be needed to induce mating. Moon cycles might have to be simulated.

Cohabitation: Specimens of different sizes can be kept together without cannibalism.

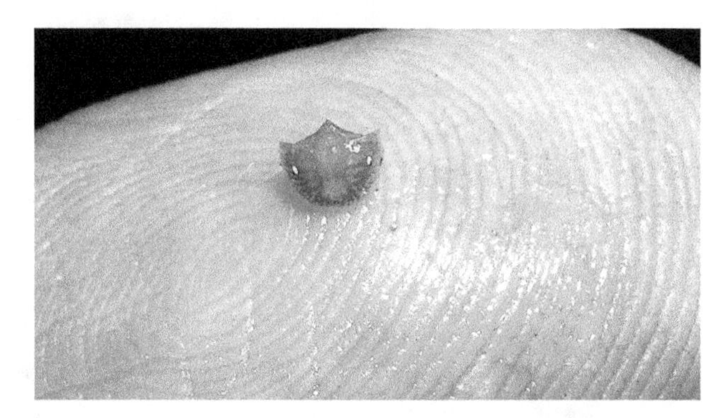

Hatchling *Limulus polyphemus* are gigantic compared to the tiny planktonic larvae of many marine arthropods. This hatchling is in a defensive position where it folds the carapace and abdomen together to protect the legs and book gills.

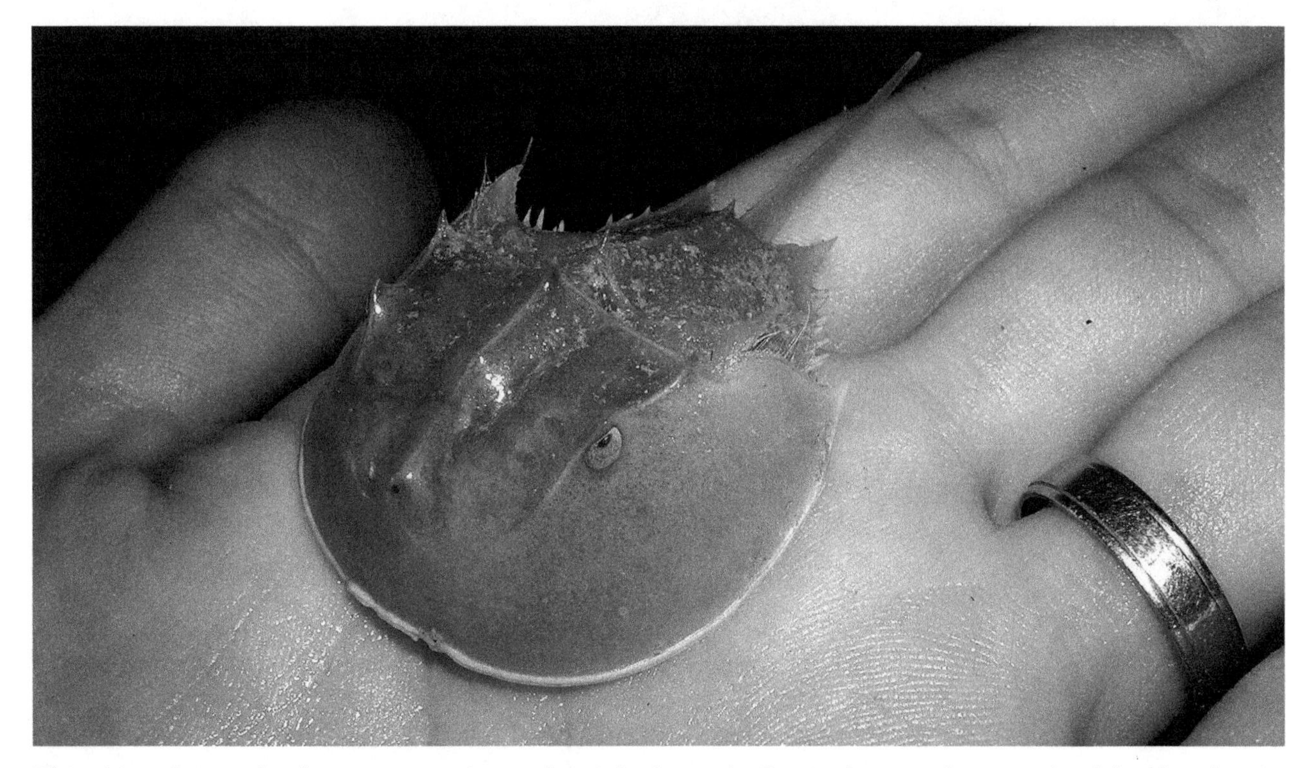

This *Limulus polyphemus* specimen kept in in a dedicated aquarium molted half a dozen times in two and a half years.

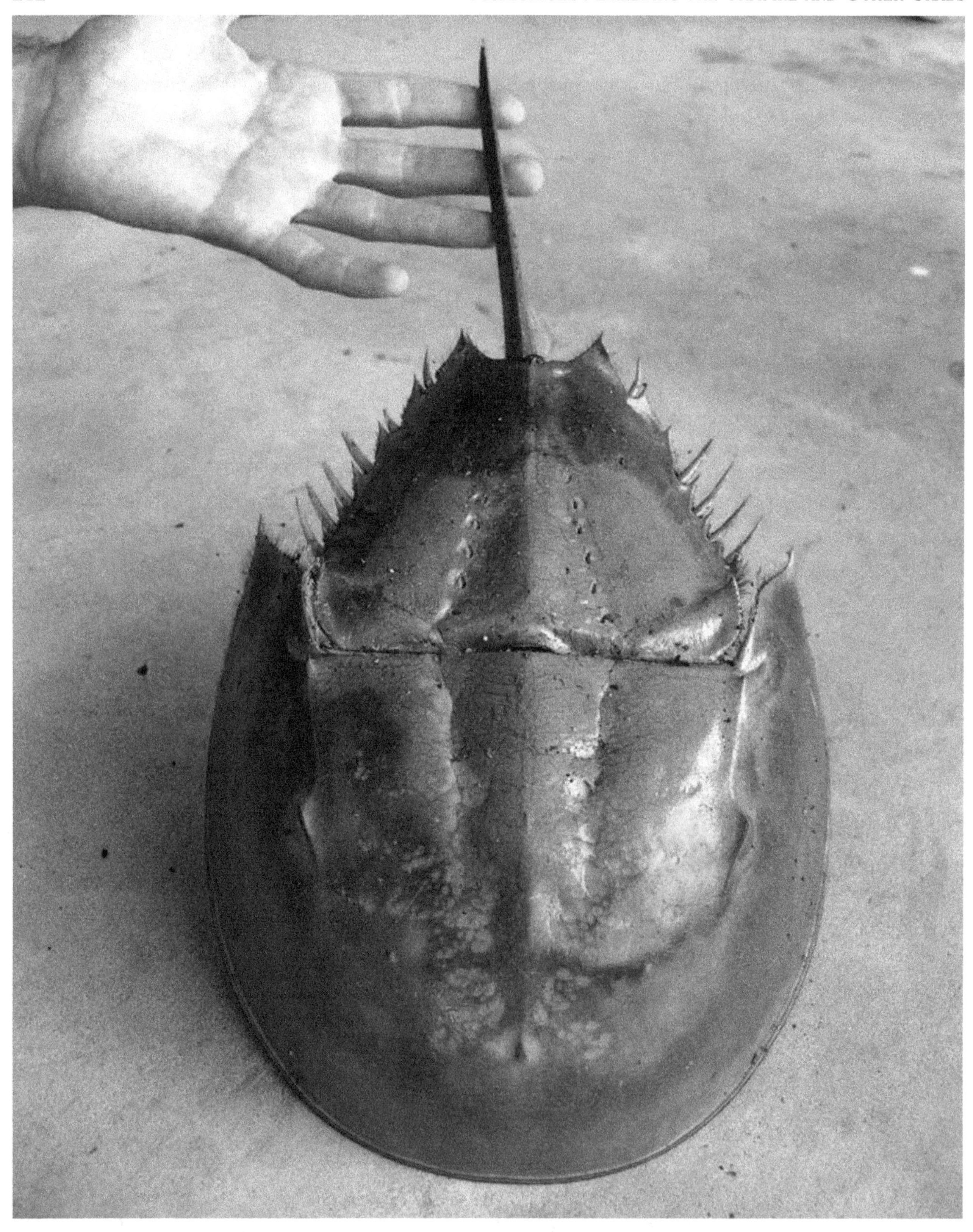

After a decade or more of growth, American horseshoe crabs can become very large. © Erin Russell

Mangrove horseshoe crab, *Carcinoscorpius rotundicauda*, Malaysia (Bernard Dupont)

Mating pair of mangrove horseshoe crabs, Hong Kong (Kevin Laurle)

The Japanese shore crab (*Hemigrapsus sanguineus*) is originally from the western Pacific but was introduced by man to eastern North America, currently found from Maine to North Carolina. Specimens accept a wide range of temperature and salinity and may get as large as a moon crab.

GLOSSARY

Anaerobic: Oxygen deficient.

Antennae: Outer pair of long, thin, unsegmented, sensory organs on the head of a crustacean whose musculature is at the base only; second antennae.

Antennules: Inner pair of crustacean sensory organs that usually move at two or more articulated points; first antennae.

Autotomize (verb): To voluntarily release an appendage, commonly in effort to escape a predator. Autotomy (noun:) The reflexive release of an appendage at a specific regrowth point.

Chela (*pl.* Chelae): The pincer formed by the last two leg segments.

Cheliped: The entire leg, ending in a chela. For crabs this is only the front pair, but other crustaceans can have multiple pairs of chelipeds.

Chitin: A fibrous polysaccharide forming the exoskeleton of arthropods and cells walls of fungi.

Coxa: Basal leg segment closest to the body (segment that remains after a leg is autotomized).

Cryptic (adj.): Well-hidden or difficult to see in a habitat. Crypsis (noun): Mimicry of common background objects like rocks or plants rather than fauna. Many crabs look like rocks or dirt (including immature *Geosesarma*), while others have setae on the exoskeleton used to hold small pieces of surrounding materials.

Dimorphic: Possessing two distinct forms or shapes. Sexually dimorphic males and females are visibly different in shape. Big differences are often expressed in the chelae, pleopods, and abdomens of crabs.

Exuvium: The shed exoskeleton.

Frass: The solid waste of invertebrates.

Gonopod: The male's secondary sexual organs used to transfer sperm; the enlarged, first pair of pleopods of male brachyurans.

Gonopore: Genital pores on a female invertebrate; located on the basal segment of the third pair of pereopods on most crabs.

Instar: Developmental stages after hatching and between molts (i.e. a third instar crab would have molted twice after hatching). Crabs' instars are rarely considered because there are so many and they vary by stage according to the species.

Littoral: Of or living within the area between high and low tide.

Macroalga (*pl.* Macroalgae): Large algae with leaf-like or root-like structures that resemble structures of vascular plants; seaweed.

Maxilliped: Outer three pairs of mouthparts that arise from the thorax. These are important tools for moving water to the gills and the reason crabs breathe through their mouths

An unidentified ghost crab (Ocypodinae) from Tortugero, Costa Rica © Robin Runck

Soldier crabs, Fraser Island, Australia © Brian Scantlebury

(unlike most arthropods that breath through apertures in the legs or body walls). The outer pair has appendages for cleaning the eyes and antenna of aquatic crabs.

MEGALOPA (*pl.* MEGALOPAE): Predatory second, and final, stage of true crab planktonic larvae. It can last a number of instars and swims using abdominal appendages.

NAUPLIUS (*pl.* NAUPLII): The first stage of hatching crustaceans, having one eye and an unsegmented body. Malacostracans (including crabs) normally pass this stage in the egg and hatch into zoea with multiple, defined body segments.

OMMATIDIUM (*pl.* OMMATIDIA): Structural unit or complete facet of the compound eye.

PALUDARIUM (*pl.* PALUDARIA): A terrarium that includes both terrestrial and aquatic areas, usually half and half.

PARTHENOGENESIS: Development through unfertilized eggs.

PEREON: Thorax of a crustacean.

PEREOPODS: The chelipeds and walking legs of crustaceans. These thoracic appendages are used for swimming in the larval stages. The hind pairs may be flattened for swimming or adapted to hold sponges, snail shells, etc.

PHORETIC: Describes an animal or plant that is carried by a host for dispersal purposes, but it does not feed on the host or pass through other developmental stages while on the host.

PLEON: Abdomen as separate from the pereon, contains the pleopods, telson, and uropods.

PLEOPODS: Biramous appendages on the pleon that form the gills and can be adapted in relation to mating for the males, known as swimmerets for many aquatic isopods.

SETA (*pl.* SETAE): Thin bristle- or hair-like extension of the arthropod exoskeleton.

STRIDULATION: Making noise or vibrations by rubbing body parts together or drumming legs against a surface, often used for calling mates.

TELSON: The last segment of the tail (abdomen) of a crustacean.

TENERAL: Soft, often pale or white-colored state of the arthropod exoskeleton immediately following a molt.

THORACIC: Of the thorax or middle body segment.

VENTRAL: Bottom or underside.

ZOEA: The first larval stage of most crabs. This planktonic stage can continue through multiple molts and uses the thoracic appendages to swim.

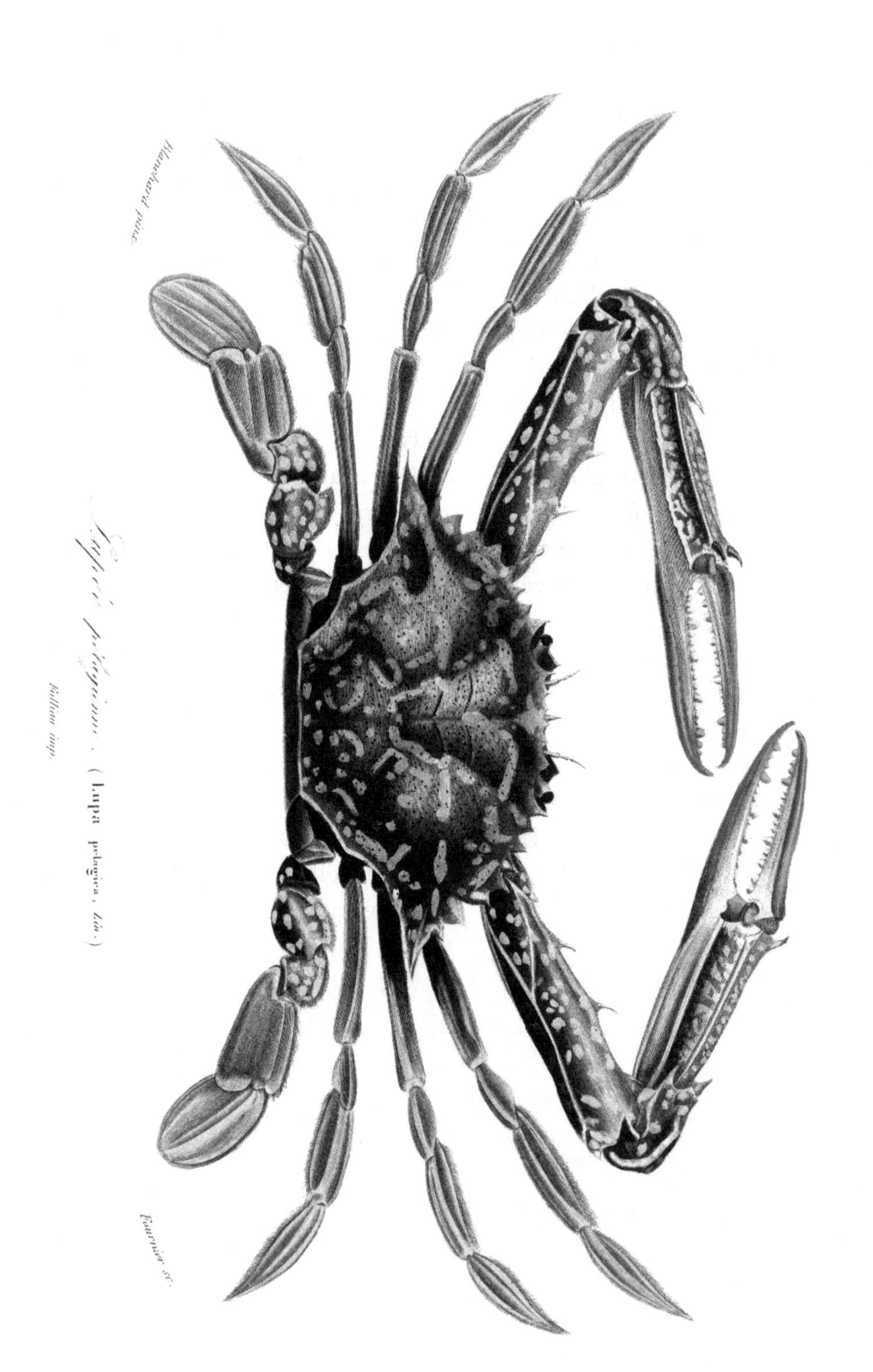
DICT. UNIV. D'HIST. NAT.
Blanchard pinx.
(Lupa pelagica, Lin.)
Bellin imp.
Fournier sc.
Crustacés. PL. 2.

BIBLIOGRAPHY

Amesbury, Steven S. (1980) *Biological Studies on the Coconut Crab (Birgus latro) in the Mariana Islands*. University of Guam Technical Report No. 17.

Botelho, E., Santos, M., and J. Souza. (2001) Aspectos populacionais do guaiamum, *Cardisoma guanhumi* Latreille, 1825, do estuário do rio Una (Pernambuco–Brasil). *Boletim Técnico Científico do Cepene* 9(1): 123-146.

Boyle, Terrence, Keith, Donald, and Russell Pfau. (2010) Occurrence, reproduction, and population genetics of the estuarine mud crab, *Rhithropanopeus Harrisii* (Gould) (Decapoda, Panopidae) in Texas freshwater reservoirs. *Crustaceana* 83(4): 493-505.

Boßelmann, F., Romano, P., Fabritius, H., Raabe, D., and M. Epple. (2007) The composition of the exoskeleton of two crustacea: The American lobster *Homarus americanus* and the edible crab *Cancer pagurus*. *Thermochimica Acta* 463(1-2): 65-68.

Butler, Henry D. (1856) *The Family Aquarium; or, a "New Pleasure" for the Domestic Circle; Being a Familiar and Complete Instructor Upon the Subject of the Construction, Fitting-up, Stocking, and Maintenance of the Fluvial and Marine Aquaria, or "River and Ocean Gardens."* New York: Dick & Fitzgerald Publishers.

Cumberlidge, Neil. (1999) *The Freshwater Crabs of West Africa; Family Potamanautidae*. Paris, France: Edition IRD.

Debelius, Helmut, and Hans Baensch. (1994) *Marine Atlas 1*. Osnabrück, Germany: Mergus Publishers.

Dost, Uwe. (2009) The tangerine-head: A tree-climber from Thailand. *TFH Magazine*.

Esser, L., and N. Cumberlidge. (2008) *Parathelphusa pantherina*. The IUCN Red List of Threatened Species 2008. http://www.iucnredlist.org/ (Downloaded 19 May 2016.)

Fox, Sue. (2000) *Hermit Crabs*. Hauppauge, NY: Barron's Education Series.

Frick, Michael G. (2003) The surf crab (*Aranaeus cribrarius*): A predator and prey item of sea turtles. *Marine Turtle Newsletter* 99: 16-18.

Fritzsche, Ingo. (2008) *Syntripsa mantannensis* (Schenkel, 1902) Krabbe aus Sulawesi. *Arthropoda* 15(3): 32.

Gosner, Kenneth. (1978) *A Field Guide to the Atlantic Seashore: Invertebrates and Seaweeds of the Atlantic Coast from the Bay of Fundy to Cape Hatteras*. Boston, Massachusetts: Houghton Mifflin Co.

Guinot, D. (2011) The position of the Hymenosomatidae MacLeay, 1838, within the Brachyura. *Zootaxa* 2890: 40–52.

Halton, Alex. (2013) *Fiddler Crabs & Fiddler Crab Care: The Complete Guide*. UK: ROC.

Mangrove tree crab, *Aratus pisonii*, Sanibel Island, Florida (James St. John)

Arboreal crab, *Armases americanum*, Costa Rica (Pavel Kirillov)

Hernáez, P., et al. (2012) Population structure and sexual maturity of the calico box crab *Hepatus epheliticus* Linnaeus (Brachyura, Hepatidae) from Yucatan Peninsula, Mexico. *Latin American Journal of Aquatic Research* 40: 480-486.

Höhle, Martin, and Martin Singheiser. (2016) *Vampirkrabben—Die Gattung Geosesarma.* Frankfurt am Main, Germany: Chimaira.

Ingle, R. (1982) "Crustacea." in Walls, Jerry G., *Encyclopedia of Marine Invertebrates.* Neptune, NJ: TFH Publications.

Järvi, Jukka (2009) Reproducción del cangrejo *Pseudosesarma moeschi.* Helsinki, Finland: Author's diary.

Kamio, Michiya, and Charles D. Derby. (2010) Chapter 20. Approaches to a molecular identification of sex pheromones in blue crabs. *Chemical Communications in Crustaceans.* pp. 393-412.

Kaplan, E. H. (1988) *A Field Guide to Southeastern and Caribbean Seashores: Cape Hatteras to the Gulf Coast, Florida, and the Caribbean.* Boston, MA: Houghton Mifflin.

Lewis, Richard C. (2007). Cape salt marsh decline linked to native crab. *The Boston Globe* (Nov. 19.)

McMonigle, Orin. (1989) *Building a "Mini-Reef" by Trial and Error.* Experimental Report, Willetts. (Hand-drawn cover includes a fiddler crab and horseshoe crab and text details include *Camposcia retusa.*)

McMonigle, Orin. (2010) D. C. invert zoo and associated microfauna search, part 2. *Invertebrates-Magazine* 10(1): 5-13.

McMonigle, Orin. (2011) *Invertebrates for Exhibition: Insects, Arachnids, and Other Invertebrates Suitable for Display in Classrooms, Museums, and Insect Zoos.* Landisville, PA: Coachwhip Publications.

McMonigle, Orin. (2014) Featured invertebrate: Thai micro crab *Limnopilos naiyanetri* Chuang & Ng 1991. *Invertebrates-Magazine* 13(3): 17.

McMonigle, Orin. (2016) Breeding the disco vampire crab, part 1. *Invertebrates Magazine* 16(1): 12-14.

McMonigle, Orin. (2017a) Breeding the disco vampire crab, part 2. *Invertebrates Magazine* 16(2): 14, 16-17.

McMonigle, Orin. (2017b) Minor complexities of the *Uca* incubator. *Invertebrates Magazine* 16(3): 16-17.

Mellen, Ida M., and Robert J. Lanier. (1935) *1001 Questions Answered About Your Aquarium.* New York: Dodd, Mead & Company.

Nash, Paul J. (1976) *Land Hermit Crabs.* Neptune, New Jersey: TFH Publications.

Ng, Peter K. L., Guinot, Daniele, and Peter J. F. Davie. (2008) Systema Brachyurorum: Part I. An annotated checklist of extant Brachyuran crabs of the world. *Raffles Bulletin of Zoology* (Issue 17).

Ng, Peter K. L., Schubart, Christoph D., and Christian Lukhaup. (2015) New species of "vampire crabs" (*Geosesarma* De Man, 1892) from central Java, Indonesia, and the identity of *Sesarma* (*Geosesarma*) *nodulifera* De Man, 1892 (Crustacea, Brachyura, Thoracotremata, Sesarmidae). *Raffles Bulletin of Zoology* 63: 3-13.

Pinheiro, Marcelo A. A., and Adilson Fransozo. (1993) Relative growth of the speckled swimming crab *Arenaeus Cribrarius* (Lamarck, 1818) (Brachyura, Portunidae), ... *Crustaceana* 65(3): 377-389.

Pronek, Neal. (1982) *Hermit Crabs.* Neptune, NJ: TFH Publications.

Rademacher, Monika, and Oliver Mengedoht. (2011) *Krabben–Fibel.* Ettlingen, Germany: Dahne Verlag GmbH.

Reich, Joe. (2010) Vampire crabs—A hobbyist's learning experiences. *Invertebrates-Magazine* 9(4): 7-10.

Rupp, Hans-Georg. (2004) The maintenance and breeding of the fiddler crab *Uca rapax* under aquarium conditions. *Aqualog News* 61: 12-15.

Ruppert, Edward, and Richard Fox. (1988) *Seashore Animals of the Southeast: A Guide to Common Shallow-Water Invertebrates of the Southeastern Atlantic Coast*. Columbia, South Carolina: University of SC Press.

Shih, Hsi-Te, et al. (2016) Species diversity of fiddler crabs, genus *Uca* Leach, 1814 (Crustacea: Ocypodidae), from Taiwan and adjacent islands, with notes on the Japanese species. *Zootaxa* 4083 (1): 57-82.

Trocini, Sabrina. (2013) Health assessment and hatching success of two Western Australian loggerhead turtle (*Caretta caretta*) populations. Thesis. Perth, Australia: Murdoch University.

Turoboyski, K. (1973) Biology and ecology of the crab *Rhithropanopeus harrisii* ssp. *tridentatus*. *Marine Biology* 23: 303-313.

Vinuesa, Julio. (2007) Molt and reproduction of the European green crab *Carcinus maenas* (Decapoda: Portunidae) in Patagonia, Argentina. *Rev. Biol. Trop.* 55: 49-54.

Vosjoli, Phillipe De. (1999) *The Care of Land Hermit Crabs*. Escondido, California: Advanced Vivarium Systems.

Walls, Jerry G. (1982) *Encyclopedia of Marine Invertebrates*. Neptune, New Jersey: TFH Publications.

Warner, G. F. (1967) The life history of the mangrove tree crab *Aratus pisonii*. *Journal of Zoology, London* 153: 321-335.

Werner, Uwe. (2003) *Aqualog Special: Shrimps, Crayfishes, and Crabs in the Freshwater Aquarium*. 2nd Edition. Rodgau, Germany: Verlag A. C. S. GmbH.

Zompro, Oliver, and Ingo Fritzsche. (2008) Neue Wirbellose–kurz vorgestellt, *Birgus latro* Linnaeus 1767–Der Palmendieb. *Arthropoda* 15(3): 26.

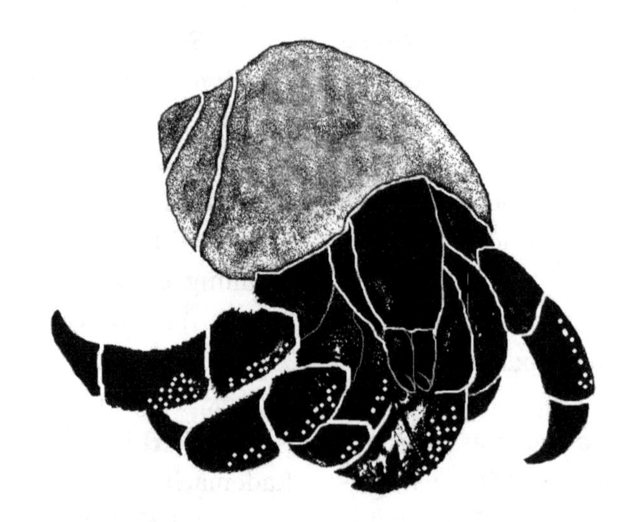

ASSASSIN BUGS,
WATERSCORPIONS,
AND OTHER HEMIPTERA
REPRODUCTIVE BIOLOGY AND
LABORATORY CULTURE METHODS
Orin McMonigle

CENTIPEDES IN
CAPTIVITY
The Reproductive Biology
and Husbandry of
Chilopoda
Orin McMonigle

BREEDING THE WORLD'S
LARGEST LIVING ARACHNID
AMBLYPYGID BIOLOGY, NATURAL HISTORY, AND CAPTIVE HUSBANDRY
ORIN MCMONIGLE

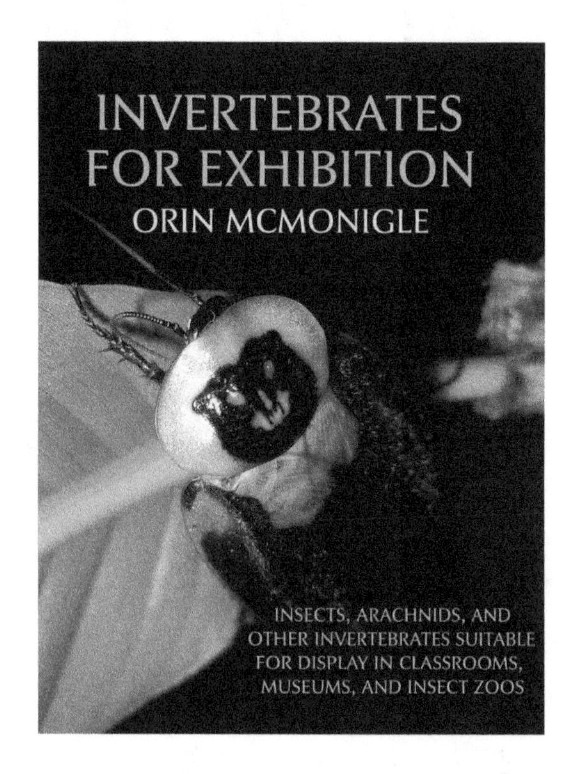
INVERTEBRATES
FOR EXHIBITION
ORIN MCMONIGLE
INSECTS, ARACHNIDS, AND
OTHER INVERTEBRATES SUITABLE
FOR DISPLAY IN CLASSROOMS,
MUSEUMS, AND INSECT ZOOS

MILLIPEDS IN CAPTIVITY
DIPLOPODAN HUSBANDRY AND
REPRODUCTIVE BIOLOGY
ORIN A. MCMONIGLE

FORGOTTEN
ORDER OF THE
VINEGAROONS
Orin McMonigle
WHIPSCORPION
BIOLOGY, HUSBANDRY,
AND NATURAL HISTORY

PILLBUGS AND
OTHER ISOPODS
CULTIVATING VIVARIUM CLEAN-UP CREWS AND
FEEDERS FOR DART FROGS, ARACHNIDS, AND INSECTS
Orin McMonigle

KEEPING THE
PRAYING MANTIS
Orin A. McMonigle

CPSIA information can be obtained
at www.ICGtesting.com
Printed in the USA
BVHW010834190919
558657BV00031BA/10/P